Lecture Notes in Computer Science   1001

Edited by G. Goos, J. Hartmanis and J. van Leeuwen

Advisory Board: W. Brauer   D. Gries   J. Stoer

Springer

Berlin
Heidelberg
New York
Barcelona
Budapest
Hong Kong
London
Milan
Paris
Santa Clara
Singapore
Tokyo

Madhu Sudan

Efficient Checking of Polynomials and Proofs and the Hardness of Approximation Problems

 Springer

Series Editors

Gerhard Goos
Universität Karlsruhe
Vincenz-Priessnitz-Straße 3, D-76128 Karlsruhe, Germany

Juris Hartmanis
Department of Computer Science, Cornell University
4130 Upson Hall, Ithaca, NY 14853, USA

Jan van Leeuwen
Department of Computer Science,Utrecht University
Padualaan 14, 3584 CH Utrecht,The Netherlands

Author

Madhu Sudan
IBM Thomas J. Watson Research Center
P.O. Box 218, Yorktown Heights, NY 10598, USA

Cataloging-in-Publication data applied for

Die Deutsche Bibliothek - CIP-Einheitsaufnahme

Sudan, Madhu:
Efficient checking of polynomials and proofs and the hardness of approximation problems /
Madhu Sudan. - Berlin ; Heidelberg ; New York ; Barcelona ; Budapest ; Hong Kong ; Lon-
don ; Milan ; Paris ; Santa Clara ; Singapore ; Tokyo : Springer, 1996
(Lecture notes in computer science ; 1001)
ISBN 3-540-60615-7

NE: GT

CR Subject Classification (1991): F.2, F.3.1, D.2.5-6, E.4, F.4.1, G.1, G.3,
I.1.2

ISBN 3-540-60615-7 Springer-Verlag Berlin Heidelberg New York

© Springer-Verlag Berlin Heidelberg 1995
Printed in Germany

Typesetting: Camera-ready by author
SPIN 10512261   06/3142 – 5 4 3 2 1 0   Printed on acid-free paper

To my dear Amma and Appa.

Foreword

How difficult is it to compute an approximate solution to an NP-optimization problem? The central importance of this issue has been recognized since the early 1970s, when Cook and Karp formulated the theory of NP-hard, and therefore computationally intractable unless P = NP, problems. To sidestep this difficulty, researchers asked whether there are polynomial time algorithms for producing near-optimal solutions to these optimization problems. This approach was successful for some problems such as bin packing, but other problems such as the Euclidean traveling salesman problem and max-clique resisted all efforts at the design of efficient approximation algorithms. Sudan's dissertation describes a general technique, akin to NP-completeness, for establishing the computational intractability of approximation problems (under the assumption that P ≠ NP). The dissertation establishes approximation hardness for all complete problems in the complexity class max-SNP: this includes basic problems such as the Euclidean traveling salesman problem, max-2SAT, and Euclidean Steiner tree. Elsewhere, these techniques have other important problems such as chromatic number, set cover, and shortest vector in a lattice. There is little doubt that the new techniques are very generally applicable, and are fundamental to establishing the intractability of approximate solutions to NP-optimization problems.

The techniques themselves are interesting and deep. They build upon a sequence of beautiful previous results on probabilistically checkable proofs. Sudan's dissertation provides a new charaterization of the complexity class NP, of languages such that membership of a string x in the language can be established by a polynomial size proof. The new characterization shows that the proofs of membership can be made surprisingly robust: the robust proofs are still polynomially long, but can be checked (in a probabilistic sense) by probing only a constant number of randomly chosen bits of the proof. The proof of this theorem is a technical tour de force; it has several major new ingredients in addition to masterfully building upon the previous work of Babai et al., Feige et al., and Arora and Safra, to name a few.

One new ingredient is a beautiful technique for creating long but very robust proofs based on self-correction properties of linear functions. Another is a new *low-degree test* that probes the value of a multivariate function at only a constant number of points and verifies whether it is close to some low-degree polynomial. The dissertation also introduces a new connection between robust probabilistically checkable proofs and the approximation hardness of the optimization problem max-SAT. This connection is the basis of the new technique for proving approximation hardness of NP optimization problems.

Sudan's dissertation introduces a new framework for the general algebraic problem of efficiently reconstructing a low degree multivariate polynomial from erroneous data. Using this framework, it presents self-contained proofs of several previous results on probabilistically checkable proofs, as well as the new results. In this framework, the connection of this work to coding theory becomes more explicit as well; the testers and correctors for multivariate polynomials developed in the dissertation yield codes with very efficient error-detection and error-correction schemes.

The work reported in this dissertation has already had, and will continue to have, a profound influence on theoretical computer science.

November 1995

Umesh Vazirani
Professor of Computer Science
University of California at Berkeley

Preface

The definition of the class NP (Cook [41], Levin [86]) highlights the problem of verification of proofs as one of central interest to theoretical computer science. Recent efforts have shown that the efficiency of the verification can be greatly improved by allowing the verifier access to random bits and accepting probabilistic guarantees from the verifier [20, 19, 50, 6]. We improve upon the efficiency of the proof systems developed above and obtain proofs which can be verified probabilistically by examining only a constant number of (randomly chosen) bits of the proof.

The efficiently verifiable proofs constructed here rely on the structural properties of low-degree polynomials. We explore the properties of these functions by examining some simple and basic questions about them. We consider questions of the form:

(**testing**) Given an oracle for a function f, is f close to a low-degree polynomial?

(**correcting**) Given an oracle for a function f that is close to a low-degree polynomial g, is it possible to efficiently reconstruct the value of g on any given input using an oracle for f?

These questions have been raised before in the context of coding theory as the problems of error-detecting and error-correcting of codes. More recently, interest in such questions has revived due to their connection with the area of program result checking. We use results from coding theory as a starting point and combine these with several algorithmic techniques including pairwise independent sampling to give efficient randomized algorithms for these tasks. As a consequence we obtain fast randomized algorithms for error-detection and error-correction for some well-known codes.

The expressive nature of low-degree polynomials suffices to capture the complexity of the class NP, and we translate our results on the efficiency of the testing and correcting procedures into two different efficiently verifiable proof systems for deciding membership questions for NP languages. One proof system generates small and somewhat efficiently verifiable proofs, and the other generates very large but very efficiently verifiable proofs. We then employ new techniques from the work of Arora and Safra [6] to compose these proof systems to obtain small proofs that can be verified by probing them in just a constant number of (randomly chosen) bits.

An important consequence of this result is that for a large variety of NP-complete optimization problems, it can be shown that finding even approximate solutions is an NP-hard problem. The particular class of optimization problems we consider is MAX SNP, introduced by Papadimitriou and Yannakakis [93]. For every MAX SNP-hard problem we show that there is a constant ϵ, such that approximating the optimum to within a relative error of ϵ is NP-hard.

This version. This version of the dissertation is essentially the same as the one filed at the University of California at Berkeley in 1992. A few proofs have been fixed to address the comments of several readers who pointed out errors in the earlier version. In addition this version has an addendum at the end of every chapter to bring the reader up to date with the various developments in the subjects covered in this thesis during the period from mid-1992 to mid-1995.

Acknowledgments

I am greatly indebted to Umesh Vazirani, my advisor, for the five years of careful nurturing that he has given me. His highly intuitive and creative approach to technical issues has left a deep impact on me. His ability to throw new light onto existing ideas have been a big factor in aiding my thinking. I owe a lot to Umesh, most importantly his taste and appreciation of mathematics, some of which has hopefully rubbed off on me. Umesh was more than just a technical adviser, and many are the times I have sought his counsel on non-technical matters, and I am thankful to him for his patience with me.

I enjoyed the numerous meetings I have had with Dick Karp, during which he monitored my progress, bolstered my confidence and at the same time provided me with his candid opinion on my work. Every meeting with Dick was a highly fulfilling experience. I'd also like to thank him for the wonderful courses he taught us, which were a great part of my learning experience at Berkeley.

A large portion of the work done here has been motivated by the work of Manuel Blum, and meetings with him over the last summer provided a turning point in my research. His enthusiasm for our work proved to be a much needed catalyst, and his pointers, which led us to the wonderful world of coding theory, were crucial to some of the results described here. Mike Luby has been a constant source of inspiration to me. He has been very generous with his time, during which I gleaned immense technical knowledge from him. His study groups at ICSI have been among my best sources of information and are largely responsible for providing me with a sound technical footing. I would like to thank Dorit Hochbaum for the many hours she spent with me in my last few months here and for sharing her wealth of knowledge on approximation algorithms with me. I would also like to thank Sachin Maheshwari at the Indian Institute of Technology at New Delhi, whose enthusiasm for the area of theoretical computer science is perhaps the single largest factor for my working in this area today.

I have been very fortunate to have found a large set of great people to work with. Ronitt Rubinfeld introduced me to the area of program checking, and much of my initial as well as current work has been done jointly with her. I am grateful to her for having shared her ideas with me, for the respect she showed me, and for encouraging me to work in this area. The results of Chapters 2 and 3 were obtained jointly with Ronitt and Peter Gemmell. I'd like to thank Rajeev Motwani for playing the role of my mentor and for passing on his keen quest for knowledge, which provided the motivation behind much of the work of Chapters 4 and 5. The work described in these chapters was done jointly with Rajeev, Sanjeev Arora, Carsten Lund and Mario Szegedy. Sanjeev deserves a special thanks for having provided me with wonderful explanations of the work he was involved in and for bringing me up to date in his area.

The academic environment at Berkeley has on the one hand been a very stimulating one and on the other it has provided me with a wonderful set of friends. Milena Mihail has been a true friend, whose advice I could always count on. I learnt much from her in my early years here. Abhijit Sahay has always provided me with a great sounding board for my ideas, as I hope I have for him. His company extended well beyond the office we shared, and, along with Savita and Usha, he has been the closest I had to a family of my own in Berkeley. I was also fortunate to have found such patient officemates in Jim Ruppert and Yiannis Emiris, who never seemed to tire of me. Thanks especially to Diane Hernek, whose company was always a pleasure. Studying at Berkeley has been a fun-filled experience and I'd like to thank Will Evans, Sridhar Rajagopalan, Sigal Ar, Dana Randall, and Z Sweedyk for this.

I was fortunate to have a company at home as stimulating as at the office: Sushil Verma, my housemate for over three years, was a great friend who always provided a willing ear to the problems I was working on. Pratap Khedkar was an infinite source of information to be tapped, and many are the times when his information helped me in my work by either setting me on the right track or stopping me from spending time on dead ends. I'd like to thank K.J. Singh, Narendra Shenoy, Audumbar Padgaonkar, Anant Jhingran, Sharad Malik, Savita Sahay, Diane Bailey, Sampath Vedant, Huzur Saran, and Rajiv Murgai for all the wonderful times.

This thesis had the (mis)fortune of being read by a lot more readers than originally planned for – and consequently many mistakes were brought out from the earlier version. I thank Mihir Bellare, Michael Goldman, Oded Goldreich, Shafi Goldwasser, Jaikumar Radhakrishnan, and Karl-Heinz Schmidt for their comments on this thesis and related issues.

Last, I would like to thank my family members – my parents and my sister Satya who have been most loving and understanding to me especially during my long absence from home; and Madhulika for entering my life. Amma and Appa, this thesis is dedicated to you.

Table of Contents

1. Introduction

The concept of a proof whose correctness can be verified in polynomial time is central to theoretical computer science. This is the defining property of the fundamental complexity class NP. Recently this notion has been extended by allowing the polynomial time verifier access to random bits and extending the notion of a proof to allow the verifier a tiny probability of accepting a fallacious proof [13, 70, 29, 53, 19, 6]. Such probabilistically checkable proofs are unexpectedly powerful and their power has been explored in several recent papers [20, 19, 50, 6]. These papers show that even verifiers with severely constrained access to the proof can check proofs of very general statements – namely proofs of membership for any NP language.

In this dissertation we carry this process to completion by showing that verifiers that access only constant number of bits in the proof can still verify membership proofs for NP languages. To do this, we have to develop some tools which reveal some new characteristics of low-degree polynomials over finite fields. Our motivation for studying these problems came from the theory of self-testing/correcting programs, due to Blum, Luby and Rubinfeld [36] and Rubinfeld [100]. It turns out that there is a fundamental connection between the testing and correcting of polynomials and the existence of efficient probabilistically checkable proofs. Here we have tried to highlight this connection be deriving previously known results as well as some of the new results in a uniform manner from our results on the testing of polynomials. [1]

The early papers on NP-completeness (Cook [41], Levin [86] and Karp [80]) linked the notion of proof verification to a variety of optimization problems. This link was used to establish the hardness of finding *exact* solutions to a variety of optimization problems, by showing them to be NP-complete. The new notions of proofs are very robust, in that even an approximately good proof would be sufficient to convince verifier of the truth of a statement. Thus, one would hope that the new proof systems should lead to hardness results for finding even *approximate* solutions to some optimization problem. Feige, Goldwasser, Lovasz Safra and Szegedy [50] were the first to bring out such a connection that shows that approximating the clique size in graphs

---

[1] The development of the proofs as described here is quite different from the way in which these results evolved. Therefore the ideas from these past developments cannot be fully localized within our exposition. An effort has been made though to provide references to the past work whose ideas are used in every step.

is hard. Inspired by this result, we bring out a different connection which enables us to show a variety of problems are hard to approximate. The problems we consider are based on the class MAX SNP, defined by Papadimitriou and Yannakakis [93]. Our result shows that for every MAX SNP-hard problem there exists a constant ϵ such that estimating the optimum value to the problem to within relative error ϵ is NP-hard.

1.1 Some problems related to polynomials

Many of the techniques used in the area of interactive proofs (probabilistically checkable proofs) are based on the properties of low-degree polynomials. The structural properties of low-degree polynomials has found wide applications in the area of coding theory. These applications are essentially based on the following observation: "The value of a univariate degree d polynomial at *any* point can be reconstructed from the value of the polynomial at any $d + 1$ places." This in turn implies that two distinct polynomials disagree with each other on most inputs, and such a statement holds even for multivariate polynomials. Next we introduce a norm for the distance between functions, which captures this fact tersely (For the sake of uniformity in our discussion we only talk of polynomials over finite fields):

Definition 1.1.1. *Let F be a finite field and let f and g be functions from the space F^m to F (i.e., f and g are functions on m-variables). The* **distance** *between f and g, denoted $d(f, g)$, is the fraction of inputs from F^m on which f and g disagree. f and g are said to be ϵ-close if $d(f, g) \leq \epsilon$.*

Using the above notation, we can express the distance property of polynomials as follows:

Lemma 1.1.1 (Polynomial Distance Lemma (cf. [104])). *Let f and g be polynomials over F in m variables with total degree at most d. Then $d(f, g) \geq 1 - \frac{d}{|F|}$.*

Thus for sufficiently large finite fields ($|F| \gg d$), the distance between two polynomials is nearly 1. Now suppose we are given a function f which is known to be very close to a degree d polynomial. The Polynomial Distance Lemma guarantees that the nearest polynomial is unique and thus can be recovered. We examine the efficiency of a reconstruction procedure by posing the following question:

Problem 1.1.1 (correcting polynomials). Let $f : F^m \to F$ be ϵ-close to some polynomial g of total degree at most d. Given an oracle for f and a setting $b_1, \ldots, b_m \in F$ to the m variables, find $g(b_1, \ldots, b_m)$.

It turns out that the univariate version of this problem is well-studied in coding theory. In Chapter 2 we describe how these methods imply an efficient randomized algorithm to solve the univariate version of this problem for any $\epsilon < 1/2$ (provided F is large enough). The running time of this algorithm is polynomial in d and $\frac{1}{(1/2-\epsilon)}$. With respect to the multivariate correction problem we show how to reduce this to the univariate case by structured random sampling from the multivariate input space. The solution so obtained for the multivariate problem solves the correction problem for any $\epsilon < 1/2$ (provided F is sufficiently large). The running time in this case is poly$(d, m, \frac{1}{(1/2-\epsilon)})$. The solution provided here can be applied back to problems in coding theory to get codes with very interesting randomized error-correcting schemes. The error-correcting scheme can find the value of any bit of the codeword using very few probes into a corrupted received word.

The next problem we consider is closely related to the problem of correcting polynomials. This problem considers the task of testing if an arbitrary function is close to some polynomial. Formally:

Problem 1.1.2 (low-degree testing). Let $f : F^m \to F$ be given by an oracle. Determine if there exists a polynomial g of degree at most d which is ϵ-close to f.

While the statements of Problems 1.1.1 and 1.1.2 are syntactically close to each other, the techniques used to solve the testing question are much more complex. The difference in the two questions may be summarized as follows: The correcting question deals only with polynomials and functions that are close to polynomials and hence uses properties satisfied by polynomials. By contrast, the testing problem could be handed *any arbitrary function* and therefore to construct testers one needs to isolate properties which are satisfied *exclusively* by polynomials (and functions close to polynomials). In Chapter 3 we provide efficient randomized solutions for this task. The final result constructs a tester that probes f in $O(d)$ places and accepts f if it is a degree d polynomial and rejects f, with high probability, if f is not close to any degree d polynomial. The running time of this tester is poly(m, d). (Note: It can be shown that both the randomization and the approximate nature of the answer are needed to solve this problem with this efficiency.)

The basic paradigm used in the tests described in Chapter 3 is the following: We describe small neighborhoods – sets of size $O(d)$ from the input space – on which the value of any degree d polynomial is forced to satisfy some simple constraints. For any specific neighborhood one can test efficiently that the constraints do indeed hold on that neighborhood. Our choice of the neighborhoods guarantees, using standard results from algebra, that "f satisfies all the neighborhood constraints" if and only if "f is a degree d polynomial". But checking that all the neighborhoods all satisfied would take too long, and the best that can be hoped for is an estimate on the fraction of neighborhoods where the constraints are violated. (Such an estimate can be easily obtained by random sampling.) We are able to establish that for certain neighborhood structures the fraction of neighborhoods with violated constraints is closely related to the distance of f from low-degree polynomials. Therefore an efficient low-degree tester need only approximate the number of violated constraints. Our proof uses a blend of probabilistic techniques and algebra and reveal new ways of characterizing polynomials which may be of independent interest.

The results on testing also contribute to coding theory by providing new randomized error-detecting mechanisms for some well-known codes. This connection is described in Section 1.3. The mechanisms also find application to the area of program checking. Such connections are described in Section 1.2.

1.1.1 Proof verification

The testing problem posed above could be interpreted in the following way. Suppose an entity, called the *verifier*, wishes to verify that a given function f is close to some low-degree polynomial, then it can do so by probing f in $O(d)$ places. Suppose we introduce another entity, called the *prover*, who wishes to persuade the verifier that f is close to a degree d polynomial and is willing to provide additional information about f to aid the process. Then how "easy" is it to persuade the verifier of such a fact. (The terminology used here comes from the area of interactive proofs – see Babai [13], Babai and Moran [18] and Goldwasser, Micali and Rackoff [70].)

We make this setting more precise now: We will expect the prover to provide the additional information about f in the form of an oracle O. The properties required of the verifier should be:

- If f is a low-degree polynomial then there must exist an oracle which persuades the verifier all the time.
- If f is not close to any polynomial then for every oracle O' the verifier should reject f with high probability.

The parameters of the verification procedure that we will examine are:

- At how many points does the verifier probe the value of f?
- How many questions does the verifier need to ask of the oracle O?
- How large is the answer given by O to any one question?
- How large is this oracle O, i.e., how many different questions could have been potentially asked of this oracle?

Problem 1.1.3 (low-degree test with help). Let $f : F^m \to F$ be given by an oracle. Provide some auxiliary information about f in the form of an oracle O, so that a verifier can test if f is ϵ-close to a degree d polynomial by probing f and O at only a constant number of places?

The solution to the low-degree test gets easily transformed in this setting to give a solution, with the feature that O is a function from F^{2m} to $F^{O(d)}$, i.e., O is polynomially large in the representation of f and provides only small pieces of information about f on any question (much smaller than, say, specifying all the coefficients of f). This solution is also described in Chapter 3.

Thus we find that statements of a very restricted nature – "f has low-degree" – can be proved in very efficient manner by the results from low-degree testing. In order, to translate this result into results about more general statements we first define an artificial problem about polynomials: "satisfiability of polynomial construction rules". It turns out that for a certain choice of parameters this problem is NP-hard. Therefore obtaining an efficiently verifiable proof for this problem results in efficiently verifiable proofs for the membership problem for all languages in NP.

Definition 1.1.2 (polynomial construction rule). *Given an initial polynomial $g^{(0)}$, a construction rule for a sequence of degree d polynomials $g^{(1)}, \ldots, g^{(l)}$ is a set of rules $r_1, \ldots, r_l$, where r_i describes how to evaluate the polynomial $g^{(i)}$ at any point $x \in F^m$ using oracles for the previously defined polynomials $g^{(0)}, \ldots, g^{(i-1)}$. A rule r_i is specified by a uniformly generated algebraic circuit over F, of size polynomially bounded in the input space (i.e., $O(\text{poly}(|F|^m))$). l is referred to as the length of such a rule. The maximum number of oracle calls made by any rule to evaluate any input is called the width of the rule and is denoted w.*

Definition 1.1.3. *A polynomial construction rule $r_1, \ldots, r_k$ is satisfiable if there exists an initial polynomial $g^{(0)}$ such that the final polynomial $g^{(l)}$ as constructed by the rule is identically zero.*

The next problem we define looks at proofs of satisfiability of polynomial construction rules.

Problem 1.1.4 (satisfiability of polynomial construction rules).
Given a construction rule $r_1, \ldots, r_l$ of length l and width w, for degree d polynomials of width w, provide an oracle O such that a probabilistic guarantee on the satisfiability of the construction rule can be obtained by probing O at a constant number of places.

Once again we expect that if the rule is not satisfiable then the verification mechanism should reject every oracle (proof) with high probability. On the other hand for a satisfiable construction rule there must exist an oracle which is always accepted by the verification mechanism.

The result on testing specific polynomials can be used to solve the above problem using oracles of the form $O : F^{O(m)} \rightarrow F^{\text{poly}(m,d,l,w)}$. This solution is described also described in Chapter 3. This problem becomes interesting because for certain carefully chosen values of the parameters involved (i.e., $|F|, d, l, w$ and m) the polynomial construction rule satisfiability problem is an NP-complete problem and this is shown in Appendix C. In particular, the following can be shown:

– Given a 3-CNF formula ϕ on n-variables, a polynomial construction rule $r_1, \ldots, r_k$ can be computed in time poly(n) such that the construction rules are satisfiable if and only if ϕ is satisfiable. Moreover, $|F|, d, l, w$ are $O(\text{polylog } n)$ and $m = \theta(\frac{\log n}{\log\log n})$.

For such values of $|F|, d, l, w$ and m the solution described above implies that there exists an oracle O of size poly(n) whose answers to questions are $O(\text{polylog } n)$ bits long such that by asking this oracle a constant number of questions suffice to get a probabilistic guarantee on the satisfiability of a 3-CNF formula. In particular this gives proofs that the verifier needs to examine only $O(\text{polylog } n)$ bits to verify membership in NP languages, and this resembles the result of Babai, Fortnow, Levin and Szegedy [19] in its efficiency. But the locality of the way the proof is probed (i.e., in a constant number of entries which are $O(\text{polylog } n)$ bits long) allows for the proofs to be checked recursively, using the techniques in the work of Arora and Safra [6]. Repeated employment of this idea yields proofs which can be verified by making $O(\log^{(c)} n)$ probes into the proof (here $\log^{(c)}$ the log function composed c times). We also provide in Chapter 4 a new proof system that provides exponentially large proofs of satisfiability, but which can be verified by probing a constant number of bits of the proof. This proof system also uses results of low degree testing and in particular the testing of linear (degree 1) functions from the work of Blum, Luby and Rubinfeld [36]. Composing the earlier developed proof systems with the new one gives us the final result: Polynomial sized proofs of membership in NP which are verifiable by probing them in a constant number of bits.

In the rest of this chapter we introduce in greater detail the areas to which the problems described above relate and describe the effect of the solutions obtained here to these areas.

1.2 Program result checking

The notion of program result checking was initiated by Blum and Kannan [34, 35, 78] as a new approach for ensuring software reliability. The approach advocates the use of fast checks performed at runtime to obtain guarantees on the correctness of the solution obtained on specific *runs* of the program. Formally, a *Checker* is defined as follows:

Definition 1.2.1. *A* **Checker** *for a function g is a (randomized) algorithm that takes as input a program P supposedly computing g and an input x and behaves as follows:*

– *If the program does not compute g correctly on the given input, then the checker must fail (P, x) with high probability.*
– *If the program computes g correctly everywhere, then the checker must pass (P, x) with high probability.*

By not checking for the correctness of the program on *every* input, the task of checking is a potentially more tractable problem than that of formally verifying its correctness. On the other hand, such guarantees still suffice to establish that the program does not produce a wrong answer any time that it is actually used. These features make program checking an attractive alternative to some of the traditional methods for guaranteeing reliability of programs. For a more detailed study of checking and a comparison of its advantages versus other techniques for ensuring software reliability, see Kannan [78] and Rubinfeld [100].

A significant amount of success has been achieved in the task of constructing checkers and checkers are available for a wide variety of problems including sorting, linear programming and number theoretic applications [2, 35, 78]. A particular approach to the task of obtaining checkers that has met with considerable success was introduced by Blum, Luby and Rubinfeld [36]. They decompose the task of checking a program into two phases: a preprocessing phase and a runtime phase. In the preprocessing phase they *test* the correctness of the program on "randomly" chosen inputs from a carefully chosen distribution. In the runtime phase they compute the value of the function correctly on arbitrary inputs using the knowledge that the program has been tested for this distribution. The former phase is the *self-testing* phase and the latter phase is referred to as the *self-correcting* phase. The notion of self-correction was also independently introduced by Lipton [87]. We state the formal definitions next. In the following definitions we consider a function g described over a finite domain and the notation $d(P, g)$ reflects the fraction of inputs on which the program P does not compute g. *(The original definitions of Blum, Luby and Rubinfeld [36] allow for more general norms that could used to estimate the distance between P and g. Here we only define the concepts for the special case of the uniform norm, since all our results work with such a norm.)*

Definition 1.2.2 ([36]). *For $\epsilon > 0$, a ϵ-self-tester T for a function g, is a randomized algorithm which uses a program P as a black box and performs as follows:*

− *If $d(P, g) = 0$ then T outputs PASS, with high probability.*
− *If $d(P, g) \geq \epsilon$ then T outputs FAIL, with high probability.*

Definition 1.2.3 ([36],[87]). *For $\epsilon > 0$, a ϵ-self-corrector C for function g, is a randomized algorithm for computing g which uses a program P as a black box and on input x, computes $g(x)$ correctly, with high probability (over internal coin flips) if $d(P, g) \leq \epsilon$.*

Sometimes, we might not be interested in actually testing a specific function g, but rather just that P has a specific property. To formalize this notion, we introduce the notion of testers for function families.

Definition 1.2.4. *The distance between a function f and a family of functions $\mathcal{F}$, denoted $\Delta(f, \mathcal{F})$, is defined to be*

$$\min_{g \in \mathcal{F}} \{d(f, g)\}$$

Definition 1.2.5. *For $\epsilon > 0$, a ϵ-self-tester T for a function family $\mathcal{F}$, is a (randomized) algorithm which uses a program P as a black box and performs as follows:*

- *If $\Delta(P, \mathcal{F}) = 0$, then T outputs PASS(with high probability).*
- *If $\Delta(P, \mathcal{F}) \geq \epsilon$ then T outputs FAIL, (with high probability).*

It is a straightforward observation that if a function g has a self-testing/correcting pair then it has a checker. Conversely, if a function has a checker, then it has a self-tester.

The existence of self-correctors for functions is a very special structural property. The existence of a self-corrector for a function g implies that g is as hard to compute on the average as in the worst case. Such an equivalence in the worst case and average case behavior is not known for many functions and might not be true of NP-hard functions. Blum, Luby and Rubinfeld use the notion of "random self-reducibility" (introduced by Abadi, Feigenbaum and Kilian [1], see also Feigenbaum [51]) to exhibit self-correctors for a large collection of functions. This property can be observed in a number of algebraic functions. For instance, say that a function g mapping from a finite group G to a group H *linear* if for all $x, y \in G$, $g(x) + g(y) = g(x + y)$ (in other words g is a homomorphism from group G to group H). Then it can be shown that all linear functions have the random self-reducibility property (see [36]). Blum, Luby and Rubinfeld [36] cite many examples of linear functions: The Mod function, Integer Multiplication, Modular Multiplication, Integer Division, Matrix Multiplication etc. all of which have self-correctors due to this observation. Lipton [87] based on the techniques of Beaver and Feigenbaum [21] has similarly shown the random self-reducibility of multivariate polynomials and thus all multivariate polynomials have self-correctors.

While the existence of self-correctors of functions may be a rare occurrence, in the cases where they have been found, proofs of correctness have been straightforward. The construction of self-testers for functions on the other hand has been invariably a much harder task. The first class of functions for which self-testers were shown was the class of *linear* functions. Blum, Luby and Rubinfeld [36] show that there exists a function family tester for the family of linear functions. Moreover, if a linear function g is given by its values on a subset S of points from the domain, then g has a tester, provided g is uniquely determined by its values on the set S. The results of Babai, Fortnow and Lund [20] also give function family testers for the family of multilinear functions and this plays an important role in their work on showing MIP=NEXPTIME.

In the language of program checking, the basic issue explored in Chapter 2 is the largest value of ϵ for which an ϵ self-corrector exists for multivariate polynomials. The main issue considered in Chapter 3 is the construction of self-testers for the family of low-degree polynomials and the self-testing of polynomials.

A number of interesting algebraic computation tasks can be expressed as computations of low-degree polynomials, e.g., the determinant of a matrix, the permanent, the product of two matrices, inverse of a matrix etc. The results of Chapters 2 and 3 can be used to construct program checkers for such problems.

1.3 Connections with coding theory

The notions of testing and correcting relate to notions of error-detection and error-correction in coding theory in a very strong sense. In this section, we briefly describe the basic tasks of coding theory and compare them with the task of testing and correcting functions. The special case of testing and correcting polynomials turns out to be particularly relevant since some well-known codes inherit their nice properties from the properties of polynomials. Examples of such codes are also given in this section, along with a description of how this connection helps in designing algorithms for both program correcting and error-correction in codes.

Definitions. A generalized[2] code over the alphabet Σ is a function E from $\Sigma^k \to \Sigma^m$, where elements of the domain of E are the *message* words and the elements in the image of E form the *codewords*. For any two n-alphabet words A and B over Σ, the *absolute distance* between A and B, is the number of places where A and B differ, and the *relative distance* (or simply *distance*) is the fraction of indices on which A and B differ. The *minimum absolute distance* of a code E is the minimum, over all possible pairs of distinct codewords in E, of the absolute distance between the two words. Similarly the *minimum relative distance* (or simply *minimum distance*) is the minimum over all possible pairs of distinct codewords in E, of the relative distance between the two words.

The computation of $E(a)$ given a message a is the task of *encoding*. Given a word $A \in \Sigma^m$, determining if it is a valid codeword of E (i.e., if there exists an a such that $A = E(a)$) is referred to as *error-detection*. Given a word $A \in \Sigma^k$, computing the codeword $E(a)$ with minimum distance from A is the task of *error-correction*. Typically this task is equivalent to the task of

---

[2] The word generalized is used to represent the fact that the codewords are not necessarily over a binary alphabet. Codes over the binary alphabet are of greater direct interest since most communication lines do transmit binary information. Nevertheless, even for the task of constructing good binary codes, constructions of good generalized codes are very useful.

computing the message a which minimizes the distance between $E(a)$ and A. The latter task is referred to as *decoding*.

Connection with testing and correcting of functions. The equivalence between testing/correcting and coding theory maybe viewed as follows. Let $\mathcal{F}$ be a function family where each function in $\mathcal{F}$ maps from a finite domain D to a range R. The functions in $\mathcal{F}$ represent the messages. The encoding of a message f is the string from $R^{|D|}$ obtained by writing the value of f explicitly on all its inputs. The task of testing membership in the function family $\mathcal{F}$, becomes the task of error-detection. The task of self-correcting a function $f' : D \to R$, becomes the task of error-correction.

The above equivalence shows that if the task of testing/correcting for some family $\mathcal{F}$ is easy, then a good code can be found for the related message class. Conversely, the existence of good codes would imply testing/correcting for a related function family. In the specific case of low-degree polynomials, the Polynomial Distance Lemma guarantees large distance between any two message words.

Next we give examples of codes which may be obtained from polynomials.

Reed Solomon Codes. Let F be a finite field of size n. The Generalized Reed Solomon codes (see [99]) encode messages from the space F^k into codewords in the range F^n as follows. Let $< c_0, \ldots, c_{k-1} > \in F^k$ be a message. The message represents the polynomial $C : F \to F$ given by $C(x) = \sum_{i=1}^{k} c_i x^i$. The encoding of the message is the sequence $\{C(x)\}_{x \in F}$. Thus the codeword is a word from F^n. Since two degree $k-1$ polynomials can agree at a maximum of $k-1$ points, the absolute distance between two codewords C and D is $n - k + 1$.

Hadamard Codes. The Hadamard codes are binary codes which map from $\{0,1\}^k$ to $\{0,1\}^{2^{k-1}}$. The construction of such codes is usually defined recursively, but here we describe it in terms of polynomials over the field GF(2). A message $< c_0, \ldots, c_{k-1} >$ represents the linear function C in $k-1$ variables $x_1, \ldots, x_{k-1}$ as follows:

$$C(x_1, \ldots, x_{k-1}) = c_0 + \sum_{i=1}^{k-1} c_i * x_i$$

The encoding of the message is the sequence

$$\{C(x_1, \ldots, x_{k-1})\}_{< x_1, \ldots, x_{k-1} > \in \{0,1\}^{k-1}}.$$

Since the codeword is a polynomial of degree 1 over a field of size 2, we observe that by the Polynomial Distance Lemma, two distinct codewords differ on at least half the places, implying that the minimum absolute distance of the Hadamard codes is 2^{k-2}.

Polynomial Evaluation Codes. Both the codes given above can be unified as follows: Let $N(m, d)$ be the number of terms (monomials) in m variables of total degree at most d. [3] Then for a finite field F of size at least d, the (m, d) polynomial code has $F^{N(m,d)}$ as the message space and $F^{|F|^m}$ as the range space. A message represents the coefficients of the $N(m, d)$ monomials, and thus a polynomial of total degree at most d. The encoding of a message consists of representing the polynomial explicitly by its value on the $|F|^m$ possible inputs. The Polynomial Distance Lemma implies that the minimum absolute distance of such codes is $|F|^m(1 - \frac{d}{|F|})$. Notice that the Reed Solomon codes are obtained by setting $m = 1$ and $d = k - 1$, and the Hadamard codes by setting $m = k - 1$ and $d = 1$.

Polynomial Extension Codes. Based on how the message is as the representation of a polynomial, we get two different kinds of coding schemes. The polynomial evaluation codes were obtained by interpreting the message as a set of coefficients. If instead, we let the message specify a polynomial by its value on a selected set of places, then we obtain the polynomial extension codes. For a subset $H \subset F$, where the cardinality of H is h, the (m, h) extension code has a message space F^{h^m} and codewords are from $F^{|F|^m}$. The message represents the value of a polynomial g of degree h in each of the m variables (and hence of total degree at most mh) at the h^m points in H^m. Such a polynomial does exist and is unique. The encoding is obtained by evaluating this polynomial at all the points in F^m. Once again the Polynomial Distance Lemma guarantees that the minimum absolute distance of this code is $|F|^m(1 - \frac{mh}{|F|})$. The advantage of specifying a code in this way is that the message is embedded in the codeword. This property turns out to be useful in many applications.

Algorithmic implications. The algorithmic implication of this connection works in both directions. In Chapter 2, it is shown how to use the techniques developed from error-correction of Reed Solomon Codes to get self-correctors for programs that compute univariate polynomials. In the other direction, the testers and correctors developed for multivariate polynomials (in Chapters 2 and 3) show that the Polynomial Evaluation Codes have extremely fast randomized error-detection and error-correction schemes. Such a randomized error-detector would guarantee that a word A is very "close" to a valid codeword, with high probability, after looking at A in very few places. Similarly, a randomized error-corrector would compute the symbol of the nearest codeword at any one location correctly, with high probability, by examining a corrupted codeword at only a few locations. Such efficient error-detecting and correcting schemes were not known prior to the work on program checking. Babai, Fortnow, Levin and Szegedy [19] were the first to use this connection to build such fast error-correcting and detecting schemes

---

[3] It is a simple combinatorial exercise to show $N(m, d) = \binom{m+d}{d}$.

for some codes. Our results improve on the efficiency of such schemes and extend it to include all the codes here.

1.4 Probabilistic checking of proofs

The notion of *interactive proofs* was introduced by Goldwasser, Micali and Rackoff [70] and Babai and Moran [13, 18]. They study languages which permit interactive proofs of membership Which are verifiable by a probabilistic verifier in polynomial time and call the collection of such languages IP[4]. Goldwasser, Micali and Wigderson [68] provided evidence to show that the class IP strictly contains NP, by showing that graph non-isomorphism, a problem not known to be in NP, can be proved efficiently interactively and thus lies in the class IP. Recent breakthroughs completely characterize the power of IP and the results of Lund, Fortnow, Karloff and Nisan [91] and Shamir [105] shows IP = PSPACE.

A related model of interactive proofs which is of more interest to us is the model where the verifier gets to ask questions from more than one non-interacting provers Ben-Or, Goldwasser, Kilian and Wigderson [29] or equivalently when the prover is assumed to be a non-adaptive entity i.e., an oracle (see the work of Fortnow, Rompel and Sipser [53]). Languages which admit efficient proofs of membership under the multiple prover proof system are said to be in the class MIP and the recent result of Babai, Fortnow and Lund [20] provides an exact characterization of this class i.e., MIP = NEXPTIME.

One way to view an oracle-based interactive proof is as follows: If we explicitly write down the answer of the oracle on every question then we get a exponential sized table which is a proof of membership in NEXPTIME languages which can be verified very efficiently (in polynomial time) by a probabilistic verifier with random access to this table. This interpretation inspired Babai, Fortnow, Levin and Szegedy [19] to define the notion of *transparent proofs*: Informally, a transparent proof of a statement of the form $x \in L$ either proves a correct statement or mistakes will appear in the proof almost everywhere, thus enabling a probabilistic verifier to spot it by a cursory examination. To formalize this concept, Babai et al. placed bounds on the running time of the probabilistic verifier and considered the kind of languages which have transparent proofs that could be verified in time $t(n)$. They scale down the result in the work of Babai, Fortnow and Lund [20] to show that all languages in NP have polynomial sized transparent proofs that can be verified in $O(\text{polylog}\, n)$ time, under the assumption that the input was presented in some error-correcting code. Such an assumption is necessary since

---

[4] The notion introduced by Babai and Moran [13, 18] is slightly different from that of Goldwasser, Micali and Rackoff [70] and goes under the name of *Arthur-Merlin* games. We shall not go into the distinctions here – the interested reader is referred to one of several surveys that have appeared on this subject [15, 65, 66].

the in $O(\text{polylog}\, n)$ time the verifier cannot even read the whole input. Notice that even under the assumption that the input is presented in an convenient form, $\Omega(\log n)$ is a lower bound on the running time of the verifier, since the verifier needs to have the power to access the entire proof.

Inspired by the work of Feige, Goldwasser, Lovasz, Safra and Szegedy [50], Arora and Safra [6] consider a model of proof system which they term *probabilistically checkable proofs*. This model, like the model of transparent proofs, is also based on the notion of a probabilistic verifier accessing an oracle (as in Fortnow, Rompel and Sipser [53]). However, instead of concentrating on the running time of the probabilistic verifier that verifies the proof, the new notion concentrates on the sensitivity of the verifier to the proof. They consider the number of bits of the proof that are actually read by the probabilistic verifier on any choice of random bits, and call this the query complexity of the probabilistically checkable proof. This parameter has no inherent logarithmic lower bounds in contrast to the running time of the verifier. Moreover, by not placing a polylogarithmic restriction on the running time of the verifier, the new notion does not require inputs to be encoded in any form. Feige, Goldwasser, Lovasz, Safra and Szegedy [50] show that every language in NP has probabilistically checkable proofs with query complexity at most $O(\log n \log\log n)$. Arora and Safra [6] improve this to show that all languages in NP have probabilistically checkable proofs with query complexity $O(\text{polyloglog}\, n)$ (for inputs of size n).

Based on this notion of a proof system Arora and Safra [6] define a class of languages PCP, with two parameters studied by Feige, Goldwasser, Lovasz, Safra and Szegedy [50]: the number of random bits used by the verifier and the query complexity. For functions $r, q : \mathcal{Z}^+ \to \mathcal{Z}^+$, the class $\text{PCP}(r(n), q(n))$ consists of all languages which have probabilistically checkable proofs where the verifier uses $r(n)$ bits of randomness and reads $q(n)$ bits of the proof to verify proofs of membership in the language. In the new terminology of Arora and Safra, the previous results may be stated as $\text{NEXPTIME} = \text{PCP}(\text{poly}(n), \text{poly}(n))$ (Babai, Fortnow and Lund [20]), $\text{NP} \subset \text{PCP}(\text{polylog}\, n, \text{polylog}\, n)$ (Babai, Fortnow, Levin and Szegedy [19]), $\text{NP} \subset \text{PCP}(\log n \log\log n, \log n \log\log n)$ (Feige, Goldwasser, Lovasz, Safra and Szegedy [50]) and $\text{NP} \subset \text{PCP}(\log n, \text{polyloglog}\, n)$ (Arora and Safra [6]). The last of these provides an exact characterization of NP (since containment in the other direction follows in a straightforward manner).

In Chapter 4 we build on and improve upon the results described above to obtain a tighter characterization of NP as $\text{PCP}(\log n, O(1))$.

1.5 Hardness results for approximation problems

The areas of "proof checking" and combinatorial optimization seem quite unrelated at a first glance. Yet, in a surprising twist, Feige, Goldwasser, Lovasz, Safra and Szegedy [50] used the new results on probabilistically checkable proofs to show hardness results for approximating the clique-size. They show that unless NP $\subset$ DTIME($n^{\log\log n}$), the size of the largest clique in a graph cannot be estimated to within super-constant factors. Subsequently, by improving the performance of the probabilistically checkable proofs, Arora and Safra [6] and Arora, Lund, Motwani, Sudan and Szegedy [8] have been able to improve this to show that approximating the clique size to within n^ϵ (for some positive ϵ) is NP-hard.

Intuitively, the connection between the probabilistically checkable proofs and the approximation hardness results are due to the following reason. The existence of "robust" (probabilistically checkable) proofs for all languages in NP implies that the membership question for any such language can be converted to a problem which has a "gap" associated with it - namely, the gap in the probability of accepting a good proof vs. the probability of accepting a bad proof. (Here a "bad" proof represents the proof of a wrong statement, rather than, say, the proof of a correct statement with a few errors in it.) This gap can be translated via approximation preserving reductions to construct graphs with a large gap in the clique size. Approximating the clique size in such graphs suffices to decide membership for languages in NP.

In Chapter 5 we show a similar connection between approximation problems and probabilistically checkable proofs. In fact, we create an optimization problem which tries to estimate the probability of acceptance of any proof for a given statement. Almost by definition this problem turns out to be NP-hard to approximate. The structure of the problem turns out be very simple and hence can be reduced to many other optimization problems. In particular, we show that the class MAX SNP, defined by Papadimitriou and Yannakakis [93], contains this problem. A large variety of approximation problems are known to be MAX SNP-hard [93, 94, 32, 33, 43, 77] and thus the result from Chapter 5 translates into non-approximability result for all these problems unless NP = P.

2. On the resilience of polynomials

In this chapter we consider the task of correcting multivariate polynomials. We restrict our attention to this problem over large finite fields. We recall the basic notation and the problem statement next:

For functions f and g mapping from F^m to F, the distance between f and g, $d(f, g)$, is defined to be the fraction of inputs from F^m where f and g disagree. g and f are said to be ϵ-close if $d(f, g) \leq \epsilon$. For a function family $\mathcal{F}$, the notation, $\Delta(f, \mathcal{F})$, represents the distance from f to the member of $\mathcal{F}$ that is closest to f.

Correcting Polynomials

Given: *An oracle to compute $f : F^m \rightarrow F$, where f is ϵ-close to some polynomial g of total degree at most d and a setting $b_1, \ldots, b_m \in F$ to the m variables.*

Output: $g(b_1, \ldots, b_m)$.

The particular parameter we will be interested in is the "resilience" of multivariate polynomials, i.e., the largest value of ϵ for which the above problem can be solved efficiently. In particular, we will be interested in solutions whose running time is $\text{poly}(m, d)$ for fixed ϵ. It is straightforward to see that when $\epsilon = 1/2$, then the above problem does not have a well-defined solution, since there might exist two polynomials g_1 and g_2 such that $d(f, g_1) = d(f, g_2) = 1/2$. Thus $\epsilon = 1/2 - \delta$ ($\delta > 0$) is the best resilience that can be attained. In this chapter we show how to attain such a resilience: In particular, we give a randomized algorithm, which runs in time $\text{poly}(d, m, \frac{1}{\delta})$, to solve the polynomial correction problem over finite fields, provided $|F| \geq \Omega((\frac{1}{\delta} + d)^2)$.

2.1 Preliminaries

We consider polynomials over a finite field F. The family of all polynomials of degree at most d on the variables $x_1, \ldots, x_m$ will be denoted $F^{(d)}[x_1, \ldots, x_m]$. A polynomial g is thus a mapping from the vector space F^m to F. We will use the vector notation x to represent an element of the domain. For $s, t \in F$, we will use the notation $s * t$ to represent their product in the field. For $t \in F$ and $h \in F^m$ the notation $t * h$ will represent the vector in F^m with each coordinate of h multiplied by t.

Definition 2.1.1. *A* **curve** *through the vector space* F^m *is a function* C : $F \rightarrow F^m$, *i.e.,* C *takes a parameter t and returns a point* $C(t) \in F^m$. *A curve is thus a collection of m functions* $c_1, \ldots, c_m$ *where each* c_i *maps elements from F to F.*

Definition 2.1.2. *If the functions* c_1 *to* c_m *can be expressed as polynomials, then the largest of the degrees of* c_i, *is defined to be the* **degree** *of the curve* C.

We will use the following fact about low-degree curves through vector spaces.

Fact 2.1.1. Let C be a curve of degree d_1 and g a polynomial on m variables of total degree d_2. Let us define g restricted to C to be the function $g|_C$: $F \rightarrow F$ where $g|_C(t) = g(C(t))$. Then g restricted to C is a polynomial of degree $d_1 d_2$.

Fact 2.1.2. Given $d+1$ points $x_1, \ldots, x_{d+1}$, from the space F^m, there exists a curve of degree d which passes through the $d + 1$ points.

Proof: This follows from the fact that one can construct degree d functions $c_1, \ldots, c_m$ such that $c_i(t_j) = (x_j)_i$ for distinct $t_1, \ldots, t_{d+1} \in F$. □

A special case of curves that will be of particular interest to us is **lines** through F^m, i.e., curves of the form $C(t) = x + t * h$. Notice that a degree d multivariate polynomial restricted to a line becomes a univariate degree d polynomial.

2.2 Achieving some resilience: random self-reducibility

The notion of random self-reducibility was introduced as a tool to implement instance-hiding schemes. The first formal definition occurs in Abadi, Feigenbaum and Kilian [1] (see also Feigenbaum [51] for a survey). Here we present a restricted definition which suffices for our purposes.

Definition 2.2.1 (random self-reducibility). *A function g mapping from a finite domain $\mathcal{D}$ to a range $\mathcal{R}$ is said to be* **random self-reducible**, *if the value of g at any input* $x \in \mathcal{D}$ *can be computed efficiently from the value of g at points* $x_1, \ldots, x_k$ *where each* x_i *is a random variable distributed uniformly over the domain $\mathcal{D}$ and the joint distribution on* $< x_1, \ldots, x_k >$ *is efficiently sampleable.*

The following is due to Blum, Luby and Rubinfeld [36].

Proposition 2.2.1 ([36]). *Every random self-reducible function has a self-corrector.*

Lipton [87] based on the work of Beaver and Feigenbaum [21] shows that the family of multivariate polynomials over large finite fields are random self-reducible.

Lemma 2.2.1 ([21],[87]). *Let $g : F^m \to F$ be a degree d polynomial, where F is a finite field such that $|F| \geq d + 2$. Then g is random self-reducible.*

Proof: Let x be any arbitrary point in F^m. Pick a point $h \in_R F^m$ and consider the "line" through the points x and $x + h$, i.e., the set of points $\{x + t * h | t \in F\}$. g restricted to this line is a univariate polynomial in t of degree at most d. Thus, for any set $S \subset F$ of size $d + 1$, we find that the value $g(x)$ can be computed (efficiently) from the values of g at $\{x + t * h | t \in S\}$ (by interpolating for the value of the univariate polynomial in t which describes g on the line $\{x + t * h | t \in F\}$ and evaluating this polynomial at $t = 0$).

Notice further, that for $t \neq 0$, $x + t * h$ is distributed uniformly over F^m. Thus if we pick S to be any subset of $F \setminus \{0\}$ of size $d + 1$, then the value of g at any fixed point x can be computed efficiently from the value of g at the $d + 1$ randomly chosen points $\{x + t * h | t \in S\}$. Such a set S exists since $|F| \geq d + 2$. $\qquad\square$

Using the above random self-reduction, the following can be shown easily.

Corollary 2.2.1. *If g is a degree d polynomial in m variables from a finite field F, then g is $\frac{1}{3(d+1)}$-resilient.*

2.3 Achieving nearly optimal resilience

In this section we consider the task of recovering from large amounts of error. For achieving this task we look at the random self-reduction of Lemma 2.2.1 more carefully. Observe, that the random self-reduction really performs as follows: It picks a univariate subdomain of F^m i.e., a line in F^m, that contains the point we are interested in, and then uses univariate interpolation on this line to find the correct value of the function at every point on the line.

Here we improve upon the above self-reduction in phases. First, we consider the restricted problem of correcting univariate polynomials and try to achieve a resilient interpolation mechanism: one that can find the value of the correct polynomial even in the presence of a significant amount of error. Next, we show how to solve the problem of multivariate self-correction, by giving a technique for picking especially "nice" univariate subdomains. The results of this section appear in [63] and [62].

2.3.1 Univariate polynomials: error correcting codes

Here, we wish to solve the following problem:

Problem 2.3.1. Given: A function $f : F \to F$ such that $\Delta(f, F^{(d)}[x]) \le 1/2 - \delta$ and a point $a \in F$.
Output: $g(a)$, where $g \in F^{(d)}[x]$ and $d(f, g) \le 1/2 - \delta$.

Note that the problem is well-posed only if g is unique with respect to the given conditions, i.e., when $|F| > \frac{d}{(2\delta)}$.

Let n be a sufficiently large number (for the purposes required here, $n = \text{poly}(d, \frac{1}{\delta})$ suffices). Pick points $x_i \in F$ randomly, for $i = 1$ to n, and let $y_i = f(x_i)$. Applying Chernoff bounds, we may conclude that with high probability, the fraction of points such that $f(x_i) \ne g(x_i)$ is approximately the same from the set $\{x_1, \ldots, x_n\}$ as from the entire field F. Choosing n large enough, the number of indices i such that $y_i \ne g(x_i)$ can be made smaller than $(n - d - 1)/2$. Thus our problem reduces to the following one:

Problem 2.3.2. Input: n pairs (x_i, y_i) such that for all but k (s.t. $2k + d < n$) values of i, $y_i = g(x_i)$, for some univariate polynomial g of degree at most $2d$.
Output: g

Such a problem arises in various ways in coding theory. If the set of x_i's exhausts all the elements of the field F, then this is the problem of decoding the Reed-Solomon codes. If the x_i are of the form ω^i, such that $\omega^t = 1$, then the problem becomes one of correcting generalized BCH codes. In the general form as it is stated above (with no constraints on the forms of the x_i's), the problem can still be solved efficiently and directly due to an elegant method of Berlekamp and Welch [30]. We state their result here; their proof is included in the appendix.

Lemma 2.3.1 (univariate self-corrector: [30]). Given n points $(x_i, y_i) \in F^2$, there exists an algorithm which finds a degree d polynomial g such that $g(x_i) = y_i$ for all but k values of i, where $2k + d < n$, if such a g exists. The running time of the algorithm is polynomial in k, d and n.

As a corollary we obtain the following:

Corollary 2.3.1. The family of univariate polynomials of degree at most d, is $1/2 - \delta$-resilient, for all $\delta > 0$.

2.3.2 Multivariate polynomials: "nice" univariate curves

We now return to the main task of self-correcting multivariate polynomials from functions that are wrong almost half the time. The problem we solve here is the following: For parameters $\delta > 0$ and a positive integer d, let F be a finite field of size $\Omega((\frac{1}{\delta} + d)^2)$. Let $g : F^m \to F$ be a multivariate polynomial of degree at most d:

Problem 2.3.3. Given : f such that $d(f, g) \leq 1/2 - \delta$ and $a_1, a_2, \cdots, a_m \in F$.

Output : $g(a_1, a_2, \cdots, a_m)$.

In this section we describe a randomized reduction from Problem 2.3.3 to the univariate self-correction problem.

We construct a subdomain $D \subset F^m$ parameterized by a single variable x (i.e., the points in the domain D are given by $\{D(x)|x \in F\}$), such that D satisfies the following properties:

1. The function $g'(x) \equiv g(D(x))$, is a polynomial whose degree is $O(d)$ in x.
2. The point $a \equiv < a_1, \ldots, a_m >$ is contained in D; In fact we will ensure that $D(0) = a$.
3. With high probability, f agrees with g on approximately the same fraction of inputs from the domain D as from the domain F^m.

The three properties listed above help us as follows: The first property ensures that we are looking at univariate polynomials over the domain D, while the second property makes sure that this helps us find $g(a_1, \ldots, a_m)$. The last property ensures that we do not lose too much information about g during the process of the reduction. Properties 1 and 3 are contrasting in nature. Property 1 requires the domain D to be nicely *structured* and expects D to be a univariate curve of constant degree in F^m. indexcurve On the other hand, Property 3 is what would be expected if D were a *random sample* of F^m.

Before going on to the construction of such a domain D, we first reexamine the reduction of Beaver and Feigenbaum (see Lemma 2.2.1). Notice that their reduction does indeed construct a univariate subdomain by picking D to be a line through the space F^m. But this construction only achieves a very weak form of property 3. This is pointed out by Gemmell, Lipton, Rubinfeld, Sudan and Wigderson [63], where it is shown, using Markov's Inequality, that if f and g agree on all but ϵ fraction of the inputs from F^m, then with probability $1 - \frac{1}{k}$, f and g agree on all but $k\epsilon$ fraction of the inputs from the domain D. This also allows Gemmell et al. [63] to show that the family of multivariate polynomials is $(1/4 - \delta)$-resilient. This construction is not of much use though if ϵ is more than $1/4$, since then the probability with which f agrees with g on at least half the points from D, is less than a half.

In order to achieve the higher degree of randomness as required by property 3, we modify the construction of Lemma 2.2.1 as follows.

Pick α and β uniformly and randomly from F^m

$$\text{Let } D_{\alpha, \beta}(x) \equiv \alpha * x^2 + \beta * x + a$$

$$D_{\alpha, \beta} \equiv \{D_{\alpha, \beta}(x)|x \in F\}$$

Each coordinate of $D_{\alpha,\beta}(x)$ is a polynomial of degree 2 in x. Hence g' is a polynomial of degree at most $2d$ in x. Also $D_{\alpha,\beta}(0) = b$. Thus we see that $D_{\alpha,\beta}$ as picked above satisfies properties 1 and 2. [1]

The following claim establishes that $D_{\alpha,\beta}$ also forms a pairwise independent sample of F^m.

Claim. For a finite field F, $b_1, b_2 \in F^m$, and for distinct $x_1, x_2 \in F \setminus \{0\}$,

$$\Pr_{\alpha,\beta}[D_{\alpha,\beta}(x_1) = b_1 \text{ and } D_{\alpha,\beta}(x_2) = b_2] = \frac{1}{|F|^{2n}}$$

Proof: For each coordinate $i \in [n]$, there exists exactly one degree 2 polynomial p_i in x, such that $p_i(0) = a_i$, $p_i(x_1) = (b_1)_i$ and $p_i(x_2) = (b_2)_i$. Thus when we pick a random polynomial p_i such that $p_i(0) = b_i$ for the ith coordinate, the probability that $p_i(x_1) = (b_1)_i$ and $p_i(x_2) = (b_2)_i$, is $\frac{1}{|F|^2}$. Since the events are independent for each coordinate, we have

$$\Pr_{\alpha,\beta}[D_{\alpha,\beta}(x_1) = b_1 \text{ and } D_{\alpha,\beta}(x_2) = b_2] = \frac{1}{|F|^{2n}}$$

$\square$

The above claim establishes that any set S of the form $S \subset \{D_{\alpha,\beta}(x) | x \in F \setminus \{0\}\}$ is a pairwise independent sample of F^n. When combined with the following lemma, the above claim shows that the domain D also satisfies Property 3.

Lemma 2.3.2. *If $S \subset F^m$ is a pairwise independent sample of n elements from F^m, and if $d(f, g) \leq 1/2 - \delta$ then the probability that f agrees with g on at least $n(1/2 + \delta) - c\sqrt{n}$ points from S is at least $1 - \frac{1}{c^2}$.*

Proof [Sketch]: Let I be the indicator variable for the condition $f = g$ i.e.,

$$I(x) = \begin{cases} 1 & \text{if } f(x) = g(x) \\ 0 & \text{otherwise} \end{cases}$$

Then by the fact that Chebyshev's Inequality holds for pairwise independent variables (see, for instance, [88]) one can conclude that the expected value of I over the domain S is very close to the expected value of I over the domain F^m. More precisely, for positive c

$$\Pr\left[|E_{x \in_R S}[I(x)] - E_{x \in_R F^m}[I(x)]| \geq c/\sqrt{|S|}\right] \leq 1/c^2$$

Thus we find that g and f agree on at least $n(1/2 + \delta) - c\sqrt{n}$ points of S. $\square$

---

[1] The idea of substituting low-degree polynomials in a single variable for the different variables, is not a new one. In particular, this has been used by Beaver, Feigenbaum, Kilian and Rogaway [22], to reduce the number of oracles used in instance hiding schemes. The underlying property that they extract is similar. They use the fact that substitution by degree t-polynomials yields t-wise independent spaces.

Thus the domain D has all the three properties required of it. Thus the problem of multivariate self-correction on the domain F^m has been reduced to the task of univariate self-correction (of $g'(0)$) on the domain $D_{\alpha,\beta}$. By Lemma 2.3.1 this can be done in time polynomial in n and d for error at most $1/2 - \delta$. Thus we have the following theorem:

Theorem 2.3.1 (multivariate self-corrector [62]). *For a positive integer d and $\delta > 0$, the family of degree d polynomials in m variables over sufficiently large finite fields F ($|F| \geq \Omega((\frac{1}{\delta}+d)^2)$) is $(1/2-\delta)$-resilient. The running time of the self-corrector is polynomial in m, d and $\frac{1}{\delta}$.*

2.3.3 Simultaneous self-correction for many points

Here we consider a slight twist on the problem of self-correction and show that the techniques of the previous sections adapt easily to handle this problem.

The new problem we consider is the following: Suppose we are interested in the value of a polynomial $g : F^m \to F$ at l places – $a_1, a_2, \ldots, a_l$ – and the only information we have about g is given by a function f such that $d(f,g) \leq 1/2 - \delta$. The problem can obviously be solved by using the self-corrector of the previous section l times. But here we give a more direct procedure which can retrieve all the l values simultaneously in one step.

To achieve the reduction we construct (with some randomization) a domain D such that the following properties hold:

1. D is a univariate curve of degree $O(ld)$.
2. D passes through the points $a_1, \ldots, a_l$.
3. D forms a pairwise independent sample of F^m (except at the points $D(1), \ldots, D(l)$).

Such a domain can be constructed by picking r_1 and r_2 randomly from F^m and then letting D be the univariate curve which contains the points $a_1, a_2, \ldots, a_l, r_1, r_2$.

We now use the univariate self-corrector of Lemma 2.3.1 to find the polynomial g' which describes g on the curve D. $g'(1), \ldots, g(l)$ gives us the desired l values.

Thus we have the following lemma:

Lemma 2.3.3. *For a positive integer d and $\delta > 0$, given a function f such that $\exists g \in F^d[x_1, \ldots, x_m]$ such that $d(f,g) \leq 1/2-\delta$, the value of g at l points can be simultaneously reconstructed from f by a reduction to one univariate reconstruction problem for a polynomial of degree $O(ld)$.*

2.4 Discussion

Higher fraction of error: The reconstruction problem. If the distance $\Delta(f, F^{(d)}[x_1, \ldots, x_m])$ is larger than $1/2$ (say .9) then the self-correction problem is ill defined (since there can be two polynomials which can agree with the oracle f at .1 fraction of the inputs) and hence cannot be solved in the form it is stated. But we could redefine the problem and instead ask for *any* function $g \in F^{(d)}[x_1, \ldots, x_m]$ which satisfies $d(f, g) < .9$. This corresponds to the maximum likilhood decoding problem in coding theory. Motivated by some applications in cryptography Goldreich and Levin [67], study this problem for the case $d = 1$ and $F = GF(2)$ and give a solution to this problem with running time bounded by a polynomial in m, the number of variables. Ar, Lipton, Rubinfeld and Sudan [3] also studied this problem where they give a number of applications for this problem. They solve this problem in time polynomial in d and m, over large finite fields, under a *restricted* model of error that suffices for their applications. They also show that an extension of the methods from here can reduce the multivariate version of this problem to the univariate case, for large finite fields. A polynomial time (in d) solution for the univariate case for *general* error still remains open.

Implications for the permanent. Lipton [87] observed that the permanent of an $n \times n$ matrix is a polynomial of degree n and is hence random self-reducible (over sufficiently large finite fields). The implication of this was that unless $\#P = BPP$, even the task of computing the permanent on a large fraction of the matrices would not be tractable (i.e., not computable in randomized polynomial time). The improvements shown here now imply that computing the permanent on even $1/2 + \delta$ fraction of all $n \times n$ matrices from a large finite field is hard unless $\#P = BPP$. Improving on this work further, Feige and Lund [49], have shown that unless $\#P = \Sigma_2^P$ (which in particular implies a collapse of the polynomial hierarchy), the permanent of $n \times n$ matrices cannot be computed on even exponentially small fraction of all matrices (over large finite fields).

Random self-reduction over small fields. Another important issue is the random self-reducibility of computations over small finite fields. Of course, for a general polynomial, this would not be achievable since in general polynomials need not differ at very many places over small fields. But for special polynomials other properties of the function can be used to achieve some resilience and this is indeed the case for the permanent over $GF(3)$ (see Feigenbaum and Fortnow [52] and Babai and Fortnow [16]). The resilience shown by them is inverse polynomial in the dimension of the matrix and it is over a distribution which is not uniform. It would be interesting to improve either of the two aspects.

Addendum. In a recent work Goldreich, Rubinfeld and Sudan [69] extend the result of Goldreich and Levin [67] giving an algorithm to reconstruct polynomials which agree with an oracle on some $\epsilon = \Omega(\sqrt{d/|F|})$ fraction of the points. Their running time is exponential in the degree d but polynomial in the number of variables m and $\frac{1}{\epsilon}$.

3. Low-degree tests

In this chapter we discuss issues on testing polynomials. The first problem we consider here is:

Low-degree testing
Given: *A function $f : F^m \to F$ as an oracle, a positive integer d and real number $\epsilon > 0$.*
Output: PASS *if $f \in F^{(d)}[x_1, \ldots, x_m]$ and* FAIL *if $\Delta(f, F^{(d)}[x_1, \ldots, x_m]) > \epsilon$.*

A closely related problem to this is the following approximation problem: "Given a function f, estimate the distance $\Delta(f, F^{(d)}[x_1, \cdots, x_m])$ to within an additive factor of $\epsilon/2$". In the following sections we describe such estimators, first for the univariate case and then for the multivariate case. (The results are expressed in terms of low-degree testing but can be converted to the approximation setting.) The main parameter of interest will be the number of probes made into f by such a tester. The tester presented in Section 3.2.3 probes the oracle for f in only $O(d)$ places. Notice that $d+2$ is a lower bound on the number of queries on f, since for any $d+1$ points from F^m and any $d+1$ values from F, there exists a polynomial which takes on those values at those places, and hence no function can be rejected by the tester.

The basic outline that all the testers described in this chapter is the following: We isolate "neighborhoods" in the input space (F^m), i.e., sets of very small size from F^m, where any degree d polynomial must show some redundancy. More specifically, we isolate neighborhoods of size $O(d)$, where the value of the polynomial on $d+1$ points forces its value at the remaining points in the neighborhood. Thus each neighborhood expresses a constraint that f must satisfy if it were a degree d polynomial. We now estimate the number of neighborhoods on which the constraints are violated. This is an easy task since for any one neighborhood, testing whether the constraint is satisfied takes poly(d) steps. Transforming this estimate on the number of unsatisfied neighborhoods into an estimate on the distance of f from the family $F^{(d)}[\ldots]$ will hence occupy most of our attention from here onwards.

The second problem we consider is the following:

Testing specific polynomials:
Given: *A function $f : F^m \to F$ as an oracle; and an "implicit description" of a polynomial g.*

Output: *An estimate for $d(f, g)$ (or alternatively PASS if $f \equiv g$ and FAIL if $d(f, g) \geq \epsilon$).*

The solution to the problem will depend on what form the "implicit" description of g takes. We elaborate on two presentations of g under which testing is possible:

- g is described by its value on enough points so as to specify it uniquely: In particular, if we are given an oracle which can provide the value of g on some space I^m, where $I \subset F$ and $|I| \geq 2d$, then we can test for g very efficiently.
- g is described by a construction. Making this notion precise requires some effort and Section 3.3.2 describes the notion along with a solution on how to test g in such circumstances.

It is worth pointing out that most of the effort involved in solving the problem of testing specific polynomials is directed towards making the notions precise. Once this is done, the solutions follow in a straightforward manner using the low-degree tests.

Finally, in this chapter, we consider the two problems described above in a slightly different setting which is related to the area of interactive proofs (probabilistically checkable proofs). More specifically, we consider the difficulty of "persuading" a verifier that a function f presented by an oracle is close to the family of multivariate polynomials of degree at most d. We show that for any function f that *is* a multivariate polynomial, there is a small amount of additional information O, such that the tester, on probing f and O at only a *constant number of values*, will be convinced that f is close to a low-degree polynomial. On the other hand if f is not close to any low-degree polynomial, then for any augmenting information O' tester would detect that f is not a low-degree polynomial. Similarly we also consider the task of persuading a verifier that a function f is close to a polynomial g where g is presented by a rule for its construction.

3.1 Univariate polynomials

3.1.1 A simple test

We start by describing a very simple tester for univariate polynomials. The tester runs in time polynomial in d, and can be used to test over any finite subset of a (potentially infinite) field. The tester is described in terms of testing over a finite field F.

The test is obtained by defining all subsets of $d + 2$ points from F to be "neighborhoods". A "neighborhood constraint" enforces the fact that on the neighborhood, the function f looks like a polynomial. Lemma 3.1.1 shows that the distance we are interested in estimating, $\Delta(f, F^{(d)}[x])$, is bounded from above by the fraction of violated constraints. The **Basic Univariate Test** estimates the latter quantity by basic sampling.

---

program Basic Univariate Test
   Repeat $O(1)$ **times**
      Pick $d+2$ distinct points $x_0, \ldots, x_d, x_{d+1} \in_R F$
      Verify that f on $x_0, \ldots, x_{d+1}$ is a degree d polynomial

---

The correctness of the tester follows from the following lemma.

Lemma 3.1.1. *Given a positive integer d, a finite field F of size at least $d+2$ and a function $f : F \to F$, if f satisfies*

$$\Pr\left[\exists g \in F^{(d)}[x] \ s.t. \ g(x_i) = f(x_i)) \ \forall i \in \{0, \ldots, d+1\}\right] \geq 1 - \delta,$$

where the probability is taken over the uniform distribution over all $d+2$-tuples $< x_0, \ldots, x_{d+1} >$ of distinct elements from F, then $\Delta(f, F^{(d)}[x]) \leq \delta$.

Proof: Let g be the degree d polynomial which minimizes $d(f, g)$ and let the distance between f and g be δ'. Now fix $z_0, \ldots, z_d$ and let h be the unique degree d polynomial such that $h(z_i) = f(z_i)$, for $i \in \{0, \ldots, d\}$. By the definition of δ', we have that

$$\Pr_{x_{d+1} \in_R F}[h(x_{d+1}) = f(x_{d+1})] \leq 1 - \delta'$$

Thus

$$\Pr_{x_0, \ldots, x_{d+1}}\left[\exists p \in F^{(d)}[x] \ s.t. \ \forall i \in \{0, \ldots, d+1\}, \ p(x_i) = f(x_i)\right]$$
$$\leq \max_{z_0, \ldots, z_d} \Pr_{x_{d+1}}[\text{ poly through } z_0, \ldots, z_d \text{ also passes through } x_{d+1}]$$
$$\leq 1 - \delta'$$

$\square$

The tester above establishes that univariate testing is an easy task and can be done in polynomial time (in the degree of the polynomial). Furthermore, the tester probes f in only $O(d)$ places. Yet, the tester given above does not reveal any new or interesting properties of polynomials, and it cannot be generalized to multivariate polynomials. Moreover from the point of viewing of testing "programs that compute polynomials" it is not very useful, since it is not "different" from a program that computes the polynomial. Next we describe a different tester for univariate polynomials which is more useful to construct program checkers. It also reveals new properties of polynomials which enable us to extend it to test multivariate polynomials.

3.1.2 A test based on evenly spaced points

The tester of this section works only for fields of the form $\mathcal{Z}_p$ for a prime p. In particular, this fact is used in Lemma 3.1.2.

Definition 3.1.1. *We say that a set of points* $\{x_0, \ldots, x_n\}$ *is evenly spaced if $\exists h$ such that* $x_i = x_0 + i * h$.

The tester on this section uses evenly spaced points as neighborhoods, i.e., neighborhoods are of the form $\{x + i * h\}_{i=0}^{d+1}$. The constraints specify that f on the neighborhoods should agree with some polynomial. Lemma 3.1.3 shows that if all neighborhood constraints are met by a function g, then g is a polynomial of degree d. Theorem 3.1.1 shows that the distance of f from the family of degree d polynomials is at most twice the fraction of violated constraints, thus showing that is suffices to test on evenly spaced neighborhoods.

The following lemma shows that interpolation (testing if a neighborhood constraint is violated) is a much easier task for evenly spaced points. In fact, the interpolation can be performed without using any multiplication and this makes the tester "different" from any function evaluating the polynomial.

Lemma 3.1.2 (cf. [110] pages 86–91). *Given a positive integer d and a prime $p \geq d+2$ The points* $\{(x_i, y_i) | i \in \{0, \ldots, d+1\}; x_i = x + i * h; x_i, y_i \in \mathcal{Z}_p\}$ *lie on a degree d polynomial if and only if* $\sum_{i=0}^{d+1} \alpha_i y_i = 0$, *where* $\alpha_i = (-1)^{(i+1)} \binom{d+1}{i}$.

Proof [Sketch]: Define the functions $f^{(j)}$, $j = 0$ to $d+1$ as follows:

$$f^{(0)}(x_i) = y_i \text{ and } f^{(j)}(x_i) = f^{(j-1)}(x_i) - f^{(j-1)}(x_{i+1})$$

The function $f^{(j)}$ agrees with a degree $d - j$ polynomial if and only if $f^{(j-1)}$ agrees with a degree $d - j + 1$ polynomial. In particular this implies that $f^{(d)}$ is a constant and thus $f^{(d+1)}(x_0) = 0$, if and only if $f^{(0)}$ is a degree d polynomial. But $f^{(d+1)}(x_0) = \sum_{i=0}^{d+1} \alpha_i y_i$.   □

Note: The proof also shows that the constants α_i never need to be evaluated. Instead the summation can be calculated by evaluating all the functions $f^{(j)}$. In all, this takes $O(d^2)$ additions and subtractions, but no multiplications.

Furthermore, evenly spaced points suffice to characterize functions that are polynomials.

Lemma 3.1.3 (cf. [110] pages 86–91). *For a positive integer d and a prime $p \geq d+2$, the function $f : \mathcal{Z}_p \to \mathcal{Z}_p$ is a polynomial of degree at most d if and only if $\forall x, h \in \mathcal{Z}_p$, $\sum_{i=0}^{d+1} \alpha_i f(x + i * h) = 0$.*

Proof [Sketch]:   Lemma 3.1.2 immediately gives the implication in one direction. The other direction, i.e., $\forall x, h \in Z_p$, $\sum_{i=0}^{d+1} \alpha_i f(x + i * h) = 0 \Rightarrow f$ is a degree d polynomial follows from looking at the special case of $h = 1$. In this case the function is specified at all points in Z_p by its values at the set $\{0, \ldots, d\}$. Moreover if g is the unique polynomial which equals f on the points $\{0, \ldots, d\}$, then g equals f everywhere. $\square$

The following theorem shows that it suffices to test that the interpolation identities hold for evenly spaced points, in order to verify that a given function has low-degree.

Theorem 3.1.1. *Given a positive integer d, a prime $p \geq d+2$ and a function $f : Z_p \to Z_p$ such that*

$$\Pr_{x, h \in_R Z_p} \left[\sum_{i=0}^{d+1} \alpha_i f(x + i * h) = 0 \right] \geq 1 - \delta \text{ where } \delta \leq \frac{1}{2(d + 2)^2},$$

then there exists a function g such that $\forall x, h \in Z_p$, $\sum_{i=0}^{d+1} \alpha_i g(x + i * h) = 0$ and $d(f, g) \leq 2\delta$.

In particular, the bounds above imply that the tester resulting from this theorem would need to probe f in $O(d^3)$ places (since to verify that the test above holds with probability $1 - O(1/d^2)$ the test would need to be repeated $O(d^2)$ times, and each repetition involves probing f at $O(d)$ places). The proof of this theorem follows Lemmas 3.1.4, 3.1.5 and 3.1.6.

Define $g(x)$ to be $\mathrm{maj}_{h \in Z_p} \sum_{i=1}^{d+1} \alpha_i P(x + i * h)$.

Lemma 3.1.4. *g and f agree on more than $1 - 2\delta$ fraction of the inputs from Z_p.*

Proof: Consider the set of elements x such that $\Pr_h[f(x) = \sum_{i=1}^{d+1} \alpha_i f(x + i * h)] < 1/2$. If the fraction of such elements is more than 2δ then it contradicts the condition that $\Pr_{x, h}[\sum_{i=0}^{d+1} \alpha_i f(x + i * h) = 0] = \delta$. For all remaining elements, $f(x) = g(x)$. $\square$

In the following lemmas, we show that the function g satisfies the interpolation formula for all x, h. We do this by first showing that for all $x, g(x)$ is equal to the interpolation of f at x by most offsets h. We then use this to show that the interpolation formula is satisfied by g for all x, h.

Lemma 3.1.5. *For all $x \in Z_p$, $\Pr_h \left[g(x) = \sum_{i=1}^{d+1} \alpha_i f(x + i * h) \right] \geq 1 - 2(d+1)\delta$.*

Proof: Observe that

$$h_1, h_2 \in_R Z_p$$
$$\Rightarrow \quad x + i * h_1 \in_R Z_p \text{ and } x + j * h_2 \in_R Z_p$$

$$\Rightarrow \quad \Pr_{h_1,h_2}[f(x + i * h_1) = \sum_{j=1}^{d+1} \alpha_j f(x + i * h_1 + j * h_2)] \geq 1 - \delta$$

$$\text{and} \quad \Pr_{h_1,h_2}[f(x + j * h_2) = \sum_{i=1}^{d+1} \alpha_i f(x + i * h_1 + j * h_2)] \geq 1 - \delta$$

Combining the two we get

$$\Pr_{h_1,h_2}\left[\begin{array}{l} \sum_{i=1}^{d+1} \alpha_i f(x + i * h_1) \\ = \sum_{i=1}^{d+1} \sum_{j=1}^{d+1} \alpha_i \alpha_j f(x + i * h_1 + j * h_2) \\ = \sum_{j=1}^{d+1} \alpha_i f(x + j * h_1) \end{array}\right] \geq 1 - 2(d+1)\delta.$$

The lemma now follows from the observation that the probability that the same object is drawn twice from a set in two independent trials lower bounds the probability of drawing the most likely object in one trial. (Suppose the objects are ordered so that p_i is the probability of drawing object i, and $p_1 \geq p_2 \geq \dots$. Then the probability of drawing the same object twice is $\sum_i p_i^2 \leq \sum_i p_1 p_i = p_1$.)   □

Lemma 3.1.6. For all $x, h \in \mathcal{Z}_p$, if $\delta \leq \frac{1}{2(d+2)^2}$, then $\sum_{i=0}^{d+1} \alpha_i g(x+i*h) = 0$ (and thus g is a degree d polynomial).

Proof: Let $h_1, h_2 \in_R \mathcal{Z}_p$. Then $h_1 + i * h_2 \in_R \mathcal{Z}_p$ implying that for all $0 \leq i \leq d+1$

$$\Pr_{h_1,h_2}\left[g(x + i * h) = \sum_{j=1}^{d+1} \alpha_j f((x + i * h) + j * (h_1 + i * h_2))\right] \geq 1 - 2(d+1)\delta$$

Furthermore, we have for all $1 \leq j \leq d+1$

$$\Pr_{h_1,h_2}\left[\sum_{i=0}^{d+1} \alpha_j f((x + j * h_1) + i * (h + j * h_2)) = 0\right] \geq 1 - \delta$$

Putting these two together we get

$$\Pr_{h_1,h_2}\left[\begin{array}{l} \sum_{i=0}^{d+1} \alpha_i g(x + i * h) \\ = \sum_{j=1}^{d+1} \sum_{i=0}^{d+1} \alpha_j \alpha_i f((x+j*h_1) + i*(h+j*h_2)) \\ = 0 \end{array}\right] \geq 1 - 2\delta(d+1)^2 > 0.$$

The lemma follows since the statement in the lemma is independent of h_1, h_2, and therefore if its probability is positive, it must be 1.

By Lemma 3.1.3 g must be a polynomial of degree at most d.   □

Proof ([]): of Theorem 3.1.1] Follows from Lemmas 3.1.4 and 3.1.6.   □

This theorem can now be used to construct a tester for univariate polynomials as follows. This tester first appeared in the works of Gemmell, Lipton, Rubinfeld, Sudan and Wigderson [63] and Rubinfeld [100].

```
program Evenly-Spaced-Test
    Repeat O(d² log(1/β)) times
        Pick x,t ∈_R Z_p and verify that ∑_{i=0}^{d+1} α_i f(x + i * t) = 0
    Reject if any of the test fails
```

Theorem 3.1.2. *If the computation of a program can be expressed by a low-degree polynomial correctly on all its inputs from Z_p, then it is passed by* **Evenly-Spaced-Test**. *If the output of the program is not $O(\frac{1}{d^2})$-close to a univariate polynomial, then with probability $1 - \beta$, it is rejected by* **Evenly-Spaced-Test**.

Proof: With confidence $1 - \beta$, the program **Evenly-Spaced-Test** will find a bad neighborhood if the fraction of bad neighborhoods is greater than $O(\frac{1}{d^2})$. If the fraction is smaller then by Theorem 3.1.1 the program's computation is $O(\frac{1}{d^2})$-close to a degree d polynomial.   □

The tester given above forms a very practical program checker for programs that compute polynomials. Though the proof given here works only for Z_p, it can easily be extended to work for functions from Z_m to Z_m. With a bit more work, the same ideas can even be used to test polynomials over the reals and the integers. Such a test is not described here. Details of this tester appear in Rubinfeld and Sudan [101].

The interesting element of the tester is that it reveals that testing low-degreeness over strongly correlated samples suffices to establish low-degreeness over the whole domain. The fact that strongly correlated samples can give a lot of structure is exploited in the next section to give very simple low-degree tests for multivariate polynomials.

3.2 Multivariate polynomials

In this section we first describe a simple extension of the "Evenly Spaced Test" of the previous section which works for multivariate polynomials. Then we work on improving the efficiency of the tester (so that the number of tests that it performs becomes smaller). The efficiency is improved by first considering the special case of bivariate polynomials and then showing how to reduce the testing of general multivariate polynomial testing to testing of bivariate polynomials.

3.2.1 Extending the evenly spaced tester

It turns out that the evenly spaced tester of the previous section easily extends to multivariate polynomials in the following way. We pick vectors x and h uniformly at random from Z_p^m, and test that the interpolation identity

holds for neighborhoods of the form $\{x, x+h, x+2*h, \ldots, x+(d+1)*h\}$.
Theorem 3.1.1 can now be extended to apply to the functions on vector spaces
to give the following extension:

Theorem 3.2.1. *Given positive integers m, d and a prime p satisfying $p \geq d+2$, if $f : \mathcal{Z}_p^m \to \mathcal{Z}_p$ is a function such that*

$$\Pr_{x, h \in_R \mathcal{Z}_p^m} \left[\sum_{i=0}^{d+1} \alpha_i f(x + i * h) = 0 \right] \geq 1 - \delta$$

for some $\delta \leq \frac{1}{2(d+2)^2}$, then there exists a function $g : \mathcal{Z}_p^m \to \mathcal{Z}_p$, such that $d(g, f) \leq 2\delta$ and

$$\forall x, h \in \mathcal{Z}_p^m \sum_{i=0}^{d+1} \alpha_i g(x + i * h) = 0$$

The proof of the above fact follows from syntactic modifications to the
proof of Theorem 3.1.1 and is hence omitted here. Moreover, we later state
and prove a theorem (Theorem 3.2.5) which subsumes this theorem.

It still remains to be shown that the function obtained from Theorem 3.2.1
is a polynomial of degree d and we include of a proof of this statement next.
The proof here only works for fields of prime order $p > md$. An improved
version appears in the full paper of Rubinfeld and Sudan [102] which yields
a proof for the case $p \geq 2d+2$. In more recent work this characterization has
been tightened to its best possible form by Friedl and Sudan [55] who show
such a theorem for $p \geq d+2$. Their result also describes tight characterizations
for fields of non-prime order.

Lemma 3.2.1. *Given positive integers m, d and a prime $p > md$, if $g : \mathcal{Z}_p^m \to \mathcal{Z}_p$ is a function such that*

$$\forall x, h \in \mathcal{Z}_p^m \sum_{i=0}^{d+1} \alpha_i g(x + i * h) = 0$$

then g is a polynomial in the m variables of total degree at most d, provided $p \geq md$.

Proof [Sketch]: We break the proof into two parts. First we show that the
function g is a polynomial of individual degree at most d in each variable.
We then show that the total degree of this low-degree polynomial is at most
d.

We first observe that by restricting our attention to h's of the form
δ_j (the vector whose coordinates are zero in all but the jth coordinate,
where it is one), we can establish that for every restriction of the values
of $v_1, \ldots, v_{j-1}, v_{j+1}, \ldots, v_m$, the function g is a degree d polynomial in v_j

(by Lemma 3.1.2). Since this holds for all $j \in \{1, \ldots, m\}$, g must be a multivariate polynomial in the variables $v_1, \ldots, v_m$ of individual degree at most d in each variable.

At this point we already have a loose bound on the total degree of g. It is at most md, since there are m variables and the degree in each variable is at most d. Now observe that a random instantiation of the type $v_j = x_j + i * h_j$, would leave g as a function of i and for $1 - \frac{md}{p}$ random choices of x and h, this would be a degree k polynomial in i where k is the total degree of g. But we know from the interpolation identity satisfied by g that every instantiation leaves it to be a degree d polynomial in i. Thus k, the total degree of g, must be d.   □

Thus Theorem 3.2.1 allows us to use the following program as a tester for multivariate polynomials.

---

program Evenly-Spaced Multivariate Test
    Repeat $O(d^2 \log(1/\beta))$ times
        Pick x and h uniformly and randomly from $\mathcal{Z}_p^m$
        Verify that $\sum_{i=0}^{d+1} f(x + i * h) = 0$
    Reject if the test fails

Theorem 3.2.2. *If the computation of a program can be expressed by a low-degree polynomial correctly on all its inputs from $\mathcal{Z}_p^m$, then it is passed by* **Evenly-Spaced Multivariate Test***. If the output of the program is not $O(\frac{1}{d^2})$-close to a polynomial, then with probability $1 - \beta$, it is rejected by* **Evenly-Spaced Multivariate Test***.*

3.2.2 Efficient testing of bivariate polynomials

In this subsection we consider the task of testing a bivariate polynomial in the variables x and y. To be more precise we consider the task of testing the family of functions whose individual degree in x and y is at most d each. (*Note that this is in contrast to the rest of this chapter (thesis?) where we usually consider only the total degree. The reason for this deviation becomes clear in Section 3.2.3.*) We will also be considering functions over some arbitrary finite field F (not necessarily of the form $\mathcal{Z}_p$) whose size will need to be sufficiently large (and will be established later).

We attempt to solve this problem by extending the **Basic Univariate Tester** of Section 3.1.1. We define some notation first:

The set of points $\{(x_0, y) | y \in F\}$ will be called the *row* through x_0. The set of points $\{(x, y_0) | x \in F\}$ will be called the *column* through y_0.

Definition 3.2.1. *For a function $f : F^2 \to F$ and a row through x_0 (column through y_0) the* **row** (**column**) **polynomial**, *is the univariate polynomial*

$r_{x_0}^{(f,d)}$ $(c_{y_0}^{(f,d)})$ *of degree d which agrees with f on the most points on the row (column). Ties may be broken arbitrarily.*

The neighborhoods for this test consists of all sets of $d + 2$ points from a single row or from a single column. Theorem 3.2.3 shows that the distance of f from a bivariate polynomial of degree d is within a constant multiplicative factor of the number of violated constraints.

---

program Basic Bivariate Test
   Repeat k times
      Pick $x_0, \ldots, x_{d+1}$; $y \in_R F$
      Verify that $\exists p \in F^{(d)}[x]$ s.t.
          for all $i \in \{0, \ldots, d+1\}$, $p(x_i) = f(x_i, y)$.
      Pick $y_0, \ldots, y_{d+1}$ $x \in_R F$
      Verify that $\exists p \in F^{(d)}[y]$ s.t.
          for all $i \in \{0, \ldots, d+1\}$, $p(y_i) = f(x, y_i)$.
   Reject if the test fails

The following theorem is the central element in the proof of correctness of the tester and appears as the Matrix Transposition Lemma in Rubinfeld and Sudan [101]. Notice that in the above test, the number of iterations has not been specified yet. This will depend on how strong our theorem is. We will establish here that $k = O(d)$ suffices. A tighter theorem on this was proved more recently by Arora and Safra [6]. They make very elegant use of the Berlekamp-Welch technique to show that $k = O(1)$ suffices! Since the efficiency of this result becomes important in the next section, we include a statement of their result in this section.

Theorem 3.2.3. *Given a positive integer d, a real number $\epsilon < 1/12d$ and a finite field F with $|F| \geq \max\{3d(\frac{1}{1-12\epsilon d}), 50d\}$, if $f : F^2 \to F$ is a function such that*

   1. For at least $1 - \epsilon$ fraction of the x's, $d(f(x, \cdot), r_x^{f,d}) \leq .49$.
   2. For at least $1 - \epsilon$ fraction of the y's, $d(f(\cdot, y), c_y^{f,d}) \leq .49$.

then there exists a polynomial g of degree at most d in x and y such that $d(f, g) \leq 4\epsilon$.

Proof [Sketch]: Call a row *good*, if the function $f_{x_0}(y) \equiv f(x_0, y)$ satisfies $\Delta(f_{x_0}, F^d[y]) \leq .49$. Since $|F| \geq 50d$, we have that for every good row (column) there exists a unique polynomial of degree at most d which agrees with the row in .51 fraction of the row. Good columns are defined similarly. Call a point *good* if it lies on a polynomial describing 0.51 fraction of the points on its row and on a polynomial describing 0.51 fraction of the points on its column. (*In particular, note that all points on bad rows and all point on bad columns are bad.*)

We first show that most points are good. We then find a $3d \times 3d$ submatrix
of points that are all good and fit a bivariate polynomial g which agrees with
f on all these points. Observe that if a row (column) has $d + 1$ good points
where f agrees with g, then f agrees with g on all the good points in its row
(column). Repeated application of this observation allows us to show that f
agrees with g on all good points.

The following can be shown by simple counting arguments to the conditions guaranteed by the theorem:

$$\Pr_{x,y \in_R F} [(x, y) \text{ is bad }] \leq 4\epsilon \tag{3.1}$$

A $3d \times 3d$ submatrix, (i.e. $X \times Y$ where $X, Y \subset F$, $|X|, |Y| = 3d$) is called
good if all the points in $X \times Y$ are good.

Claim: A good submatrix exists.

Proof: Consider a random set $X = \{x_1, \ldots, x_{3d}\}$ of the rows. The expected
number of bad points on these rows is at most $12d\epsilon|F|$. Thus, if $12d\epsilon < 1$ and
$|F| \geq 3d(\frac{1}{12\epsilon d})$, then with positive probability, at least $3d$ columns contain
no bad points in the rows indexed by X. Call these $2d$ columns the set Y.
$X \times Y$ is a good submatrix.

Since a good submatrix has the property that all points on the submatrix
lie on degree polynomials of its rows and columns, it follows that there exists
a polynomial g of degree d in x and y which agrees with the entire submatrix.
We now show that g agrees with all good points.

*Claim: If a bivariate polynomial g of degree at most d agrees with f on a
good $3d \times 3d$ submatrix, then it agrees with f on all good points.*

Proof: The basic idea behind each of the following steps is that if $d + 1$
good points from a row (column) agree with g then all good points on that
row (column) agree with g. First, observe that all good points from the rows
indexed by X agree with g. Next, observe that at least 51% of the good
columns will have more than $d + 1$ good points in the rows indexed by X,
and for these columns, all the good points agree with g. Now consider all the
good rows: At least 2% of the points from any good row are both good and
lie on one of the 51% columns selected by the previous step, and hence the
best polynomial for these rows must agree with g. Thus we have that f at
all good points agrees with g. Thus $\Pr_{x,y \in F} [f(x, y) = g(x, y)] \geq 1 - 4\epsilon$. $\square$

Theorem 3.2.4 ([6]). *There exists constants $\epsilon_0 > 0$, c such that for every
positive integer d, if $\epsilon < \epsilon_0$, and F is a finite field of order at least cd^3 and
$f : F^2 \to F$ is a function which satisfies:*

1. *The function $R : F^2 \to F$ given by $R(x, y) = r_x^{(f,d)}(y)$ satisfies $d(f, R) \leq \epsilon$.*

2. *The function $C : F^2 \to F$ given by $C(x, y) = c_y^{(f,d)}(x)$ satisfies $d(f, C) \leq \epsilon$.*

then there exists a polynomial g of degree at most d in x and y such $d(f,g) \leq 4\epsilon$.

3.2.3 Efficient reduction from multivariate polynomials to bivariate polynomials

In this section we relate the work of the previous section to the proof of Theorem 3.2.1. The connection yields improved testers for multivariate polynomials and can be used to construct testers for the family of degree d polynomials which look at the value of a function at only $O(d)$ points to test it. The proof of this section is essentially from Rubinfeld and Sudan [102]. The efficiency shown here is better than that shown in [102], due to the use of Theorem 3.2.4 which is from [6].

Definition 3.2.2. *For $x, h \in F^m$, the set of points $\{x + t * h | t \in F\}$, denoted $l_{x,h}$, is called the line through x with offset h.*

An alternate way of stating Theorem 3.2.1 is the following: "If the fraction of lines for which f restricted to the line is not close to a univariate polynomial is small $(o(\frac{1}{d^2}))$, then f is close to some multivariate polynomial g." By doing a more careful analysis of the proof of Theorem 3.2.1 this result can be improved, and this will be the focus of this section. We first need a few definitions.

Given an integer $d > 0$, and a function $f : F^m \to F$ we describe a test which uses a help function B, which maps a pair of points x, h to a polynomial p supposedly describing the best polynomial fitting the function f on the line $l_{x,h}$. The test considers a random line $l_{x,h}$ and picks a random point $x + th$ on this line and verifies that $B(x, h)[t] = f(x + th)$. We shall not be particularly concerned with how one can evaluate the best polynomial $B(x, h)$ for any line $l_{x,h}$ – we leave it to the reader to figure out such a test using the method described in Section 3.1.1. In what follows we shall simply assume that an oracle provides the function B as well as f and show how to use B to test f. We now resume our definitions.

$$\text{Let } \delta_{f,d}^{(B)}(x) \equiv \Pr_{h \in F^m, t \in F} [f(x + th) \neq B(x, h)[t]].$$

For functions $f : F^m \to F$ and $B : F^{2m} \to F^{(d)}[t]$, let

$$\delta_{f,d}^{(B)} \equiv E_{x \in F^m} \left[\delta_{f,d}^{(B)}(x) \right].$$

Lastly we define

$$\delta_{f,d} = \min_{B:F^{2m} \to F^{(d)}[t]} \left\{ \delta_{f,d}^{(B)} \right\}.$$

We now state the main theorem for this section.

Theorem 3.2.5. *There exists a positive real δ_0 and a constant c such that the following holds. Given positive integers m and d, a finite field F of order at least $\max\{cd^3, md+1\}$, and functions $f : F^m \rightarrow F$ and $B : F^{2m} \rightarrow F^{(d)}[t]$ such that $\delta_{f,d}^{(B)} \leq \delta_0$, then there exists a degree d polynomial $g : F^m \rightarrow F$ such that $d(f,g) \leq 2\delta_{f,d}^{(B)}$.*

The following definition outlines the function B which minimizes $\delta_{f,d}^{(B)}$ for any function f.

Definition 3.2.3. *For a function $f : F^m \rightarrow F$, and points $x, h \in F^m$, we define the **line polynomial** for the line $l_{x,h}$ to be the degree d polynomial $P_{x,h}^{(f,d)}$ which maximizes the number of points $t \in F$ for which $P_{x,h}^{(f,d)}[t]$ equals $f(x + th)$.*

Remark: In the above definition we allow ties to be broken arbitrarily, but consistently. I.e., if for the two pairs (x, h) and (x', h'), $l_{x,h}$ is the same set as $l_{x',h'}$, then for a point $y = x + th = x' + t'h'$, the line polynomials must agree at y (i.e., $P_{x,h}^{(f,d)}[t] = P_{x',h'}^{(f,d)}[t']$).

Proposition 3.2.1. *For a function $f : F^m \rightarrow F$ let $B^{(f,d)} : F^{2m} \rightarrow F^{(d)}[t]$ be given by $B^{(f,d)}(x, h) = P_{x,h}^{(f,d)}$. Then*

$$\delta_{f,d}^{(B^{(f,d)})} = \min_{B:F^{2m} \rightarrow F^{(d)}[t]} \left\{ \delta_{f,d}^{(B)} \right\}.$$

Proof: The proposition follows from the observation that, for any $x, h \in F^m$, the quantity

$$\min_{p \in F^{(d)}[t]} \left\{ \Pr_{t \in F}[f(x + th) \neq p(t)] \right\},$$

is minimized by the polynomial $p = P_{x,h}^{(f,d)}$. $\qquad\square$

Proposition 3.2.1 shows that it suffices to prove that if $\delta_{f,d}^{(B^{(f,d)})} \leq \delta_0$ then f is $2\delta_{f,d}^{(B^{(f,d)})}$-close to some degree d polynomial.

In what follows we will show that the following "self-corrected" version of f is close to f. For any function $f : F^m \rightarrow F$, define $\mathrm{corr}_{f,d} : F^m \rightarrow F$ to be

$$\mathrm{corr}_{f,d}(x) = \mathrm{maj}_{h \in F^m} \left\{ P_{x,h}^{(f,d)}(0) \right\}.$$

It is easy to show that f is always close to $\mathrm{corr}_{f,d}$ and we do so next.

Lemma 3.2.2. *For any function $f : F^m \rightarrow F$ and any integer d,*

$$d(f, \mathrm{corr}_{f,d}) \leq 2\delta_{f,d}.$$

Proof: Let B be the set given by

$$B = \left\{ x \in F^m \mid \Pr_{h \in F^m} [f(x) \neq P_{x,h}^{(f,d)}(0)] > 1/2 \right\}.$$

Then the probability that for a randomly chosen pair x, h that $f(x) \neq P_{x,h}^{(f,d)}(0)$ is at least $(1/2) \cdot \Pr_x[x \in B]$. But by the definition of $\delta_{f,d}$ and Proposition 3.2.1, we know that the quantity $\Pr_{x,h}[f(x) \neq P_{x,h}^{(f,d)}(0)] = \delta_{f,d}$. Thus the probability that x lies in B is at most $2\delta_{f,d}$. The lemma follows from the observation that for $x \notin B$, $\text{corr}_{f,d}(x) = f(x)$. $\qquad\square$

The bulk of the work in proving the correctness of Theorem 3.2.5 is in showing that $\text{corr}_{f,d}$ is a degree d polynomial if $\delta_{f,d}$ is sufficiently small. In order to prove this theorem for general multivariate polynomials, we use Theorem 3.2.4. Recall the definition of a row and column, and the row polynomial and column polynomial from Section 3.2.2.

We use Theorem 3.2.4 as follows:

Lemma 3.2.3. *Let c and ϵ_0 be as in Theorem 3.2.4. Given reals ϵ, ϵ', an integer d, a field F, a function $f : F^2 \to F$ and elements $x_0, y_0 \in F$ which satisfy the following properties:*

1. *$\epsilon \leq \epsilon_0$, $|F| \geq cd^3$ and $5\epsilon + \epsilon' + \frac{(d+1)}{|F|} < 1/2$.*

2. *The function $R : F^2 \to F$ given by $R(x,y) = r_x^{(f,d)}(y)$ satisfies $d(f, R) \leq \epsilon$.*

3. *The function $C : F^2 \to F$ given by $C(x,y) = c_y^{(f,d)}(x)$ satisfies $d(f, C) \leq \epsilon$.*

4. *The quantities $\Pr_x[f(x, y_0) \neq r_x^{(f,d)}(y_0)]$ and $\Pr_y[f(x_0, y) \neq c_y^{(f,d)}(x_0)]$ are at most ϵ'*

Then $r_{x_0}^{(f,d)}(y_0) = c_{y_0}^{(f,d)}(x_0)$.

Proof: By applying Theorem 3.2.4 we find that there exists a polynomial g of degree d each in x and y such that $d(f, g) \leq \epsilon$. By conditions (2) and (3), we conclude that $d(g, C), d(g, R) \leq 5\epsilon$. This allows us to say that:

$$\Pr_x[g(x, \cdot) \neq R(x, \cdot)] \leq 5\epsilon + \frac{d}{|F|} \qquad (3.2)$$

(since for values of x where $R(x, \cdot) \neq g(x, \cdot)$ contribute at least $|F|(1 - \frac{d}{|F|})$ points y such that $g(x, y) \neq R(x, y)$). Condition (4) of the lemma statement guarantees that $\Pr_x[f(x, y_0) \neq r_x^{(f,d)}(y_0)] \leq \epsilon'$. Combining this with (3.2) we find

$$\Pr_x[f(x, y_0) \neq g(x, y_0)] \leq 5\epsilon + \epsilon' + \frac{d}{|F|}.$$

Now based on the condition $5\epsilon + \epsilon' + \frac{d}{|F|} < 1/2 - \frac{d}{|F|}$, we conclude that $g(\cdot, y_0)$ is the unique degree d polynomial which describes the y_0th column and hence $c_{y_0}^{(f,d)} = g(\cdot, y_0)$. This implies $c_{y_0}^{(f,d)}(x_0) = g(x_0, y_0)$. A similar argument yields $r^{(f,d)})_{x_0}(y_0) = g(x_0, y_0)$ which yields the assertion of the lemma. $\qquad\square$

Lemma 3.2.4. *Let ϵ_0 and c be as in Theorem 3.2.4. Given a finite field F of cardinality at least $\max\{3(d+1), cd^3\}$, a function $f : F^m \to F$, and a constant $\epsilon < \min\{\frac{1}{36}, \epsilon_0\}$, we have:*

$$\forall x \in F^m, t_0 \in F, \quad \Pr_{h_1, h_2}\left[P_{x,h_1}^{(f,d)}(t_0) \neq P_{x+t_0 h_1, h_2}^{(f,d)}(0)\right] \leq \frac{4\delta_{f,d}}{\epsilon} + \frac{4}{|F|}.$$

Proof: We use the shorthand δ for $\delta_{(f,d)}$. Pick $h_1, h_2 \in_R F^m$ and let $m : F^2 \to F$ be the function given by $m(y, z) = f(x + y h_1 + z h_2)$. We will show that with probability at least $1 - 4(\delta/\epsilon + 1/|F|)$ (taken over h_1 and h_2) m satisfies the conditions required for the application of Lemma 3.2.3 for $y_0 = t_0$ and $z_0 = 0$ and $\epsilon' = \epsilon$. This will suffice to prove the lemma since $P_{x,h_1}^{(f,d)} = c_{z_0}^{(m,d)}$ and $P_{x+t_0 h_1, h_2}^{(f,d)} = r_{y_0}^{(m,d)}$ and the lemma guarantees that $c_{z_0}^{(m,d)}(y_0) = r_{y_0}^{(m,d)}(z_0)$.

We start by observing that for the chosen values of ϵ and ϵ', $5\epsilon + \epsilon' + (d+1)/|F| < 1/2$. We now go on to showing that the remaining conditions required for applying Lemma 3.2.3.

Since $x + y h_1$ and h_2 are random and independent of each other, we have

$$\forall y \neq 0, z, \quad \Pr_{h_1, h_2}\left[P_{x+y h_1, h_2}^{(f,d)}(z) \neq f(x + y h_1 + z h_2)\right] \leq \delta. \tag{3.3}$$

Notice that the event above $P_{x+y h_1, h_2}^{(f,d)}(z) \neq f(x + y h_1 + z h_2)$ maybe rephrased as $r_y^{(m,d)}(z) \neq m(y, z)$. Applying Markov's inequality to (3.3), we find that

$$\Pr_{h_1, h_2}\left[\Pr_{y \neq 0, z}[m(y, z) \neq r_y^{(m,d)}(z)] \geq \epsilon\right] \leq \frac{\delta}{\epsilon}.$$

By accounting for the probability of the event $y = 0$, we can conclude:

$$\Pr_{h_1, h_2}\left[\Pr_{y, z}[m(y, z) \neq r_y^{(m,d)}(z)] \geq \epsilon\right] \leq \frac{\delta}{\epsilon} + \frac{1}{|F|}. \tag{3.4}$$

Applying Markov's Inequality again to (3.3), but this time fixing $z = z_0$, we get

$$\Pr_{h_1, h_2}\left[\Pr_{y}[m(y, z_0) \neq r_y^{(m,d)}(z_0)] \geq \epsilon\right] \leq \frac{\delta}{\epsilon} + \frac{1}{|F|}. \tag{3.5}$$

A similar argument to the above yields

$$\Pr_{h_1, h_2}\left[\Pr_{y, z}[m(y, z) \neq c_z^{(m,d)}(y)] \geq \epsilon\right] \leq \frac{\delta}{\epsilon} + \frac{1}{|F|}. \tag{3.6}$$

and $\Pr_{h_1, h_2}\left[\Pr_{z}[m(y_0, z) \neq c_z^{(m,d)}(y_0)] \geq \epsilon\right] \leq \frac{\delta}{\epsilon} + \frac{1}{|F|}.$ $\tag{3.7}$

Thus with probability at least $1 - 4(\delta/\epsilon + 1/|F|)$ none of the events (3.4)-(3.7) happen and the conditions required for Lemma 3.2.3 are satisfied. □

An application of Markov's inequality yields the following corollary:

Corollary 3.2.1. *Let ϵ_0, c be as in Theorem 3.2.4. For any constant $\epsilon <$* $\min\{\frac{1}{36}, \epsilon_0\}$, *if F is a finite field of size at least $\max\{3(d+1), cd^3\}$ and $f :$* $F^m \to F$, *then*

$$\forall x \in F^m, t \in F, \quad \Pr_{h \in F^m}\left[\text{corr}_{f,d}(x+th) \neq P^{(f,d)}_{x,h}(t)\right] \leq \frac{8\delta}{\epsilon} + \frac{8}{|F|}.$$

Proof: Let $B_{x,t}$ be the set defined as

$$B_{x,t} = \left\{h \in F^m | P^{(f,d)}_{x,h}(t) \neq \text{maj}_{h'}\{P^{(f,d)}_{x+th'}(0)\}\right\}.$$

For $h \in B_{x,t}$, the probability that for a randomly chosen h_1 that $P^{(f,d)}_{x+th,h_1}(0) \neq$ $P_{x,h}(t)$ is at least $1/2$. Thus with probability at least $\frac{|B_{x,t}|}{2|F|^m}$, we find that a randomly chosen pair h, h_1, violates the condition $P^{(f,d)}_{x+th,h_1}(0) = P_{x,h}(t)$. Applying Lemma 3.2.4 we get that $\frac{|B|}{|F|^m}$ is at most $2 \cdot (4\delta/\epsilon + 4/|F|)$. On the other hand, we have that for all $h \notin B$, $\text{corr}_{f,d}(x+th) = P^{(f,d)}_{x,h}(t)$. $\qquad\square$

Even the specialization of the lemma above to the case $t = 0$ is particularly interesting since it says that the "majority" in the definition of $\text{corr}_{f,d}$ is actually well-defined and an overwhelming majority, provided $\delta_{f,d}$ is sufficiently small. The next lemma essentially shows that $\text{corr}_{f,d}$ is a degree d polynomial.

Lemma 3.2.5. *Let ϵ_0 and c be as in Theorem 3.2.4. Let F be a finite field of size at least $\max\{3(d+1), cd^3\}$. For $\epsilon = \min 1/36, \epsilon_0$ let $f : F^m \to F$ be such that $\delta = \delta_{f,d}$ satisfies*

$$\frac{256\delta}{\epsilon^2} + \frac{256}{\epsilon|F|} + \frac{56\delta}{\epsilon} + \frac{40}{|F|} < 1.$$

Then $\quad \forall x, h \in F^m \quad \text{corr}_{f,d}(x) = P^{(\text{corr}_{f,d},d)}_{x,h}(0)$

Proof: Pick $h_1, h_2 \in_R F^m$ and define $M : F^2 \to F$ to be

$$M(y, 0) = \text{corr}_{f,d}(x + yh) \text{ and } M(y,z) = f(x + yh + zh_1 + yh_2)) \text{ for } z \neq 0.$$

We will show, by an invocation of Lemma 3.2.3, that the event $c_0^{(M,d)}(0) =$ $r_0^{(M,d)}(0)$ happens with probability strictly greater than $8\delta/\epsilon + 8/|F|$ over the random choices of h_1 and h_2. But by Corollary 3.2.1, we have $M(0,0) =$ $\text{corr}_{f,d}(x) = P^{(f,d)}_{x,h_1}(0) = c_0^{(M,d)}(0)$ with probability at least $1 - 8\delta/\epsilon - 8/|F|$. (Of the three equalities in the chain here — the first and the third are by definition and the middle one uses Corollary 3.2.1.) This would suffice to prove the lemma since we will have established that with positive probability (over the choice of h_1 and h_2), $\text{corr}_{f,d}(x) = r_0^{(M,d)}(0) = P^{(g,d)}_{x,h}(0)$. But the event stated is invariant with h_1 and h_2 and hence its probability, if positive, must be 1. Thus it remains to show that the conditions required for Lemma 3.2.3 are true for the function M, with $\epsilon' = \epsilon$ and $y_0 = z_0 = 0$.

For $z = 0$ and all y, we have $M(y,z) = \mathrm{corr}_{f,d}(x + yh + z(h_1 + yh_2))$, by definition. For any $z \neq 0$ and y, the quantity $\Pr_{h_1,h_2}[M(y,z) \neq \mathrm{corr}_{f,d}(x + yh + z(h_1 + yh_2))]$ is at most 2δ (by Lemma 3.2.2). Also, for all $y, z \in F$, the probability that $\mathrm{corr}_{f,d}(x + yh + z(h_1 + yh_2))$ does not equal $P^{f,d}_{x+yh,h_1+yh_2}(z)$ is at most $8\delta/\epsilon + 8/|F|$ (by Corollary 3.2.1). Thus we find

$$\Pr_{h_1,h_2}\left[M(y,z) \neq c_z^{(M,d)}(y)\right] \leq 8\delta/\epsilon + 8/|F| + 2\delta = \delta_1.$$

As in the proof of Lemma 3.2.4 we can now conclude that

$$\Pr_{h_1,h_2}\left[\Pr_{y,z}[M(y,z) \neq c_z^{(M,d)}(y)] \geq \epsilon\right] \leq \frac{\delta_1}{\epsilon} + \frac{1}{|F|}. \tag{3.8}$$

$$\text{and} \quad \Pr_{h_1,h_2}\left[\Pr_z[M(0,z) \neq c_z^{(M,d)}(0)] \geq \epsilon\right] \leq \frac{\delta_1}{\epsilon} + \frac{1}{|F|}. \tag{3.9}$$

For the rows the conditions required are shown even more easily. We first observe that the line $l_{x+zh_1,h+zh_2}$ is a random line through F^m for any $z \neq 0$. Thus we can use the definition of δ to claim that

$$\forall y \in F, \quad \Pr_{h_1,h_2}\left[\begin{array}{l} M(y,z) \\ = f(x + zh_1 + y(h + zh_2)) \\ = P^{(f,d)}_{x+zh_1,h+zh_2}(y) \\ = r_z^{(M,d)}(y) \end{array}\right] = 1 - \delta.$$

As in the proof of Lemma 3.2.4 we can argue that

$$\Pr_{h_1,h_2}\left[\Pr_{y,z}[M(y,z) \neq r_y^{(M,d)}(z)] \geq \epsilon\right] \leq \frac{\delta}{\epsilon} + \frac{1}{|F|}. \tag{3.10}$$

$$\text{and} \quad \Pr_{h_1,h_2}\left[\Pr_z[M(0,z) \neq r_y^{(M,d)}(0)] \geq \epsilon\right] \leq \frac{\delta}{\epsilon} + \frac{1}{|F|}. \tag{3.11}$$

Thus with probability at least $1 - 16(\delta_1/\epsilon + \delta/\epsilon + 2/|F|)$ none of the events (3.8)-(3.11) happen and we can apply Lemma 3.2.3.

To conclude we need to show that $1 - 16(\delta_1/\epsilon + \delta/\epsilon + 2/|F|) > 8\delta/\epsilon + 8/|F|$ and this follows from the conditions given is the statement of the lemma. $\square$

Proof ([): of Theorem 3.2.5] Let $\epsilon = \min 1/36, \epsilon_0$, where ϵ_0 is as in Theorem 3.2.4. Now pick α to be $\max\{c, 600/\epsilon\}$, where c is as given by Theorem 3.2.4. Now let $\delta_0 = \frac{\epsilon^2}{624}$. Notice that δ_0 is positive.

For any function $f : F^m \to F$, over a field of size at least $\alpha(d+1)^3$, if $\delta_f < \delta_0$, then the conditions required for the application of Lemma 3.2.5 are true and we can conclude that $\mathrm{corr}_{f,d}$ satisfies

$$\forall x, h \quad \mathrm{corr}_{f,d}(x) = P^{(\mathrm{corr}_{f,d},d)}_{x,h}.$$

By Lemma 3.2.1 this condition is equivalent to saying $\mathrm{corr}_{f,d}$ is a degree d polynomial, provided $|F| \geq md$. By Lemma 3.2.2 $d(f, \mathrm{corr}_{f,d}) \leq 2\delta_{f,d}$. Thus f is $2\delta_{f,d}$-close to a degree d polynomial. $\square$

3.3 Testing specific polynomials

In this section we consider the task of testing if a function f is almost equal to a specific polynomial g. There are various possibilities of how g might be specified. For instance, we might know the value of g at sufficiently many places to determine it uniquely. Alternatively, g might be a well known polynomial, like the determinant or the permanent and hence we might know enough properties about g to determine it uniquely. Here we consider two cases by which the polynomial might be specified, and how to test in those cases.

3.3.1 Polynomials specified by value

Here we consider a polynomial $g : F^m \rightarrow F$ whose value on some subset of the form I^m, where $I \subset F$, is provided by an oracle.

Lemma 3.3.1. *Given an arbitrary function f and a specification of a degree d polynomial g by its values at some subset I^m of points, testing that $\mathrm{Pr}_{x \in F^m}[f(x) = g(x)] \geq 1 - \delta$ takes $O(d\frac{|I|}{|I|-d})$ probes.*

Proof: Using the tester given by Theorem 3.2.5 we first verify that there exists a multivariate polynomial g' of degree at most d such that $d(f, g') \leq \delta$. Next we make sure that g' equals g for most points in I^m. In order to do this we should be able to compute g' for any one point in I^m (notice that f might not equal g' on any point in I^m!). To do this we use the **Multivariate Self Corrector** given by Theorem 2.3.1, to compute $g'(x)$ for $x \in I^m$. Now we can estimate the quantity $\mathrm{Pr}_{x \in F^m}[g(x) = g'(x)]$. By the Polynomial Distance Lemma [104] we know that if this quantity is greater than $\frac{|I|-d}{|I|}$ then $g \equiv g'$. Thus using $O(\frac{|I|}{|I|-d})$ evaluations of g' we can test whether $g' = g$ or not. Thus using $O(d\frac{|I|}{|I|-d})$ probes into f we can test if g is close to f or not. $\square$

3.3.2 Polynomials specified by construction

Here we consider the case where the polynomial is specified by a construction. We define the notion of a construction rule for polynomials and then show that polynomials specified by construction rules are testable. We illustrate the power of such construction rules by demonstrating a simple example of such a rule for the permanent.

Definition 3.3.1 (polynomial construction rule). *Given an initial polynomial $g^{(0)}$, a construction rule for a sequence of degree d polynomials $g^{(1)}, \ldots, g^{(l)}$ is a set of rules $r_1, \ldots, r_l$, where r_i describes how to evaluate the polynomial $g^{(i)}$ at any point $x \in F^m$ using oracles for the previously defined polynomials $g^{(0)}, \ldots, g^{(i-1)}$. A rule r_i is a uniformly constructed algebraic circuit over F whose size is polynomially bounded in the input space* *size (i.e., $O(\text{poly}(|F|^m)))$. l is referred to as the length of such a rule. The maximum number of oracle calls made by any rule to evaluate any input is called the width of the rule and is denoted w.*

The tester for such functions can be obtained from the downward-self-reducibility theorem of Blum, Luby and Rubinfeld [36]. We include a version of their theorem here that is weaker than the original theorem they prove. The tester also resembles the Sum-Check protocols of Lund, Fortnow, Karloff and Nisan [91], Shamir [105] and Babai, Fortnow and Lund [20].

Let $\delta > 0$ be a constant smaller than the constant of Theorem 3.2.5.

Theorem 3.3.1 ([36]). *Given oracles for a sequence of functions $f^{(1)}, \ldots, f^{(l)}$, an oracle for an initial polynomial $g^{(0)}$, and a construction rule $r_1, \ldots, r_l$ of width w for degree d polynomials $g^{(1)}, \ldots, g^{(l)}$ ($g^{(i)} : F^m \to F$) there exists a tester which verifies that the $f^{(i)}$ is δ-close to $g^{(i)}$ for all $i \in \{1, \ldots, l\}$. Moreover the test probes the sequence f in $\text{poly}(l, w, d)$ points.*

Proof: The tester for this sequence steps through the sequence establishing inductively that $f^{(i)}$ is δ-close to $g^{(i)}$. Assume inductively that the tester has established that $f^{(i)}$ and $g^{(i)}$ are close, for $i < k$. To establish that $f^{(k)}$ and $g^{(k)}$ are close, we first establish that $f^{(k)}$ is close to a degree d polynomial and then test if $f^{(k)}(x) = g^{(k)}(x)$ for randomly chosen points $x \in F^m$. To evaluate $g^{(k)}(x)$, the tester can use rule r_k, provided it has access to oracles for $g^{(i)}$, $i < k$. To obtain an oracle for $g^{(i)}$, $i > 0$, the oracle uses the fact that $f^{(i)}$ is close to $g^{(i)}$, and thus the **Multivariate Self Corrector** given by Theorem 2.3.1 can be used to simulate an oracle for $g^{(i)}$. Thus $g^{(k)}(x)$ can be computed in time $\text{poly}(w, d)$ using oracle calls to $f^{(i)}$, $i < k$. The tester can thus test the entire sequence using $\text{poly}(l, w, d)$ calls to the oracles for the sequence f and $g^{(0)}$. □

Notice that the above proof does not really use the fact that we are working with polynomials. It only uses the random-self-reducibility of the functions $g^{(i)}$. Yet we have shown here only the weaker form of this theorem. In the next section we will improve upon the efficiency of this theorem in a related model. This improvement will be used in later chapters to obtain transparent proofs of NP-hard statements.

Note on the running time of the above tester: The running time of the above tester is effectively bounded (within factors of $\text{poly}(l, w, d)$) by the representation of the rule $r_1, \ldots, r_k$ and the time taken to evaluate such any of the r_i's.

The following example illustrates the power of such testers.

Example 3.3.1. **The Permanent:** Let $g^{(n)}(x_{11}, \cdots, x_{ij}, \cdots, x_{nn})$ be the permanent of the $n \times n$ matrix whose ijth entry is x_{ij}. Then $g^{(n)}$ is a polynomial of degree at most n. Furthermore $g^{(n)}$ can be constructed from $g^{(n-1)}$ as follows:

$$g^{(n)}(x_{11}, \ldots, x_{nn}) = \sum_{i=1}^{n} x_{i1} * f^{(n-1)}(X^{(i)}(x_{11}, \ldots, x_{nn}))$$

where the function $X^{(i)}$ projects the $n \times n$ vector so as to obtain the $(1, i)$th minor of the matrix given by $\{x_{ij}\}$.

Thus the permanent can be represented by a constructible sequence of polynomials, and hence by Theorem 3.3.1 can be tested.

Lastly we also make the following observation on the amount of randomness needed by the tests given here.

Corollary 3.3.1 (to Theorem 3.3.1). *A sequence of functions* $g^{(0)}, \ldots,$ $g^{(l)}$, $g^{(i)} : F^m \to F$, *given by a construction sequence can be tested using* $O(m \log |F| + l)$ *random bits.*

Proof: Notice that any one phase (i.e., testing is $f^{(i)}$ is close to $g^{(i)}$) takes $O(m \log |F|)$ random bits. A naive implementation of the tester would thus take $O(km \log |F|)$ random bits. But we can save on this randomness by recycling the random bits via the technique of Cohen and Wigderson [38] or Impagliazzo and Zuckerman [74]. This would allow each additional phase to reuse most of the old random bits, and would need only a constant number of fresh random bits per phase. Thus the whole algorithm can be implemented using $O(m \log |F| + k)$ random bits. □

3.4 Efficient testing of polynomials in the presence of help

The previous sections concentrated on the task of verifying that a function f is close to a polynomial. This could be viewed in the setting of interactive proofs where the verifier (tester) is being persuaded of this fact without any help from the prover. We now consider the situation where the prover is allowed to help the verifier (tester) by providing additional information about the function f. We expect the additional information to be presented in the form of an oracle O which is queried by the verifier. (In this section we will use the words "verifier" and "tester" interchangeably.) Formally:

Proofs of Low-degree
Given an oracle for a function $f : F^m \to F$, *is it possible to specify some additional information* O *about* f *so that a tester can verify that* f *is close to a degree d polynomial. In particular, the tester should reject f with high probability if f is not close to a low-degree polynomial. On the other hand, if f is a degree d polynomial there should exist an O such that the tester always accepts f.*

The following parameters will be of interest to us:

- The running time of the tester.
- The size of the oracle O.
- The length of O's response on any single question.
- The number of questions asked of O by the verifier.
- The number of probes made by the verifier into f.

Theorem 3.4.1. *There exists a tester T which accesses two oracles f and O such that given a function $f : F^m \to F$*

- *f is a polynomial of degree at most $d \Rightarrow \exists$ an oracle O such that $T(f, O)$ outputs PASS.*
- *$\Delta(f, F^{(d)}[x_1, \ldots, x_m]) \geq \delta \Rightarrow \forall$ oracles O, $T(f, O')$ outputs FAIL with high probability.*

O can be expressed as a function from F^{2m} to F^{d+1}, i.e., the size of O is quadratic in the size of the oracle for f and the length of the responses of O are $\mathrm{poly}(d, \log |F|)$ bits. Moreover, T probes f and O in only a constant number of places.

Proof: The tester that yields this theorem is given below.

```
program T(f,O);
    Repeat O(1/δ²) times
        Pick x, h ∈ᵣ Fᵐ and t ∈ᵣ F
        Let p = O(x, h)
        Verify that p(t) = f(x + t * h)
        Reject if the test fails
```

The proof of correctness follows from Theorem 3.2.5. $\qquad\square$

Lastly we consider a problem in the spirit of the problem considered in Theorem 3.3.1, which we state slightly differently to make the problem statement simpler. The motivation for this problem will become clear in the next chapter.

Recall the definition of a polynomial construction rule (Definition 3.3.1).

Definition 3.4.1. *A polynomial construction rule $r_1, \ldots, r_l$ is satisfiable if there exists a polynomial $g^{(0)}$ such that the sequence of polynomials $g^{(1)}, \ldots, g^{(l)}$, computed from $g^{(0)}$ according to the construction rule, terminates with $g^{(l)} \equiv 0$.*

The next problem we consider deals with proofs of satisfiability of construction rules.

Satisfiability of polynomial construction rules
Given a construction rule $r_1, \ldots, r_l$ of width w for degree d polynomials, provide an oracle O which proves the satisfiability of the rule. In particular, if

the rule is satisfiable, then there must exist an oracle which is always accepted by the tester. On the other hand, if the rule is not satisfiable, then any oracle O' must be rejected by the tester with high probability.

The starting point for our solution to this problem is the result of Theorem 3.3.1, i.e., the oracle can provide tables for the polynomials $g^{(0)}, \ldots, g^{(l)} \equiv 0$ and then the tester of Theorem '3.3.1 can be used to verify that the construction rules have been obeyed. In order to cut down on the number of questions asked of such an oracle we use some of the work done on parallelizing the MIP = NEXPTIME protocol by Lapidot and Shamir [85], and some improvements on it by Feige and Lovasz [48].

Theorem 3.4.2. *Given polynomial construction rule $r_1, \ldots, r_l$ of width w, for degree d polynomials from F^m to F for a sufficiently large finite field ($|F| > \mathrm{poly}(l, w, d, m)$ suffices), there exists a tester T such that:*

- *If $r_1, \ldots, r_l$ is satisfiable, then there exists an oracle O such that T always outputs PASS.*
- *If $r_1, \ldots, r_l$ is not satisfiable, then for all oracles O', the tester T outputs FAIL with high probability.*

Moreover the size of the oracle is $\mathrm{poly}(|F|^m, 2^l)$ and the response of the oracle to any question is $\mathrm{poly}(l, w, d, \log|F|, m)$ bits long and the tester probes O in only a constant number of points.

Proof: The tester here will be constructed so as to simulate the tester of Theorem 3.3.1 and the oracle will be created so as to aid this tester. The oracle will thus be expected to provide the oracles for the functions $g^{(0)}, \ldots, g^{(l)}$. An obvious simulation of the tester of Theorem 3.3.1 would involve performing low-degree tests on each of the $l + 1$ functions and hence too many probes into the oracle. Instead we will work with a single polynomial G, defined on $m + 1$ variables, which represents all of the functions $g^{(i)}$.

$$G(z, x) \equiv g^{(i)}(x) \text{ if } z \text{ is the } i\text{th element of } F \text{ for } 0 \le i \le l$$

Such a polynomial G exists with degree l in z and total degree in at most $d + l$.

Suppose the oracle O provides us with a function f supposedly representing G. We will expect the oracle to augment it with f_{lines}, which takes two points x and h as argument and returns the value of the best polynomial fitting f on the line through x with offset h. Using f_{lines} we can test according to Theorem 3.4.1 to see if there exists a low-degree polynomial $\tilde{f}$ which is δ-close to f.

Furthermore, we can compute $\tilde{f}$ at any point x by picking a random point h and using f_{lines} to find the polynomial describing h on the line through x and $x + h$. Evaluating this polynomial at 0 gives us $\tilde{f}(x)$. Evaluating this polynomial at $t \in_R F$ and cross checking with $h(x + t * h)$ ensures us that the wrong polynomial is not returned by f_{lines}. Thus effectively we have an oracle for $\tilde{f}$.

The polynomial $\tilde{f}$ gives us a sequence of polynomials $f^{(i)} : F^m \to F$ and we need to ensure that $f^{(i)} = g^{(i)}$ (i.e., the construction rules have been obeyed). The tester of Theorem 3.3.1 shows how to do this by looking at the value of the functions $f^{(i)}$ at $O(lw)$ points. In order to find the value of the functions $f^{(i)}$ on these points, which is equivalent to finding the value of $\tilde{f}$ at some lw points $x_1, \ldots, x_{lw}$, we use the idea of "simultaneous self-correction" (see Lemma 2.3.3). We construct a curve C of degree lw which passes through the points $x_1, \ldots, x_{lw}$. Observe that $\tilde{f}$ restricted to this curve must be some polynomial p of degree at most $(d + l)lw$. Now we expect the oracle to provide the value of $\tilde{f}$ restricted to the curve C (explicitly). Suppose the oracle returns a polynomial p'. If $p = p'$ then we are done, since the we can now simulate the tester of Theorem 3.3.1. But the oracle may describe a polynomial $p' \neq p$. In order to detect this we pick a random value of $t \in F$ and check that $p'(t) = \tilde{f}(C(t))$ and these two will be different with probability $1 - \frac{(d+l)lw}{|F|}$.

In summary, we have:

Oracle. The oracle consists of the functions f, f_{lines} and a function f_{curves} which describes $\tilde{f}$ on some curves of degree lw. The number of different curves that T may query about is bounded by the number of different random strings used by the tester which is $2^{O(m \log |F| + d + l)}$. Thus f_{curves} can be expressed as function from $2^{O(m \log |F| + d + l)} \to F^{(d+l)lw}$ and this dominates the size of the oracle.

Tester. Tests that f is close to a low-degree polynomial $\tilde{f}$. Then it simulates the action of the tester of Theorem 3.3.1 by generating all the points where the value of $\tilde{f}$ is required. It constructs a low-degree curve which passes through all these points and queries f_{curves} for the polynomial describing $\tilde{f}$ on this curve. It then queries f_{lines} and f at once each to reconstruct the value of $\tilde{f}$ at one point. Finally it makes sure that $f^{(l)}$ is identically zero. In all this takes a constant number of probes into O (some constant for the low-degree test, 3 for the questions needed to simulate the tester of Theorem 3.3.1 and a constant to ensure $f^{(l)} \equiv 0$). □

3.5 Discussion

History. A tester for multivariate polynomials was first constructed by Babai, Fortnow and Lund [20]. This was followed up by more efficient versions in Babai, Fortnow, Levin and Szegedy [19] Feige, Goldwasser, Lovasz, Safra and Szegedy [50] and Arora and Safra [6]. All these testers have one common ingredient: They test for the degree of each variable individually and thus have an inherent $\Omega(m)$ lower bound on the number of probes required. The tester developed in this chapter was developed in a concurrent stream of research by Gemmell, Lipton, Rubinfeld, Sudan and Wigderson [63], Shen [106] and Rubinfeld and Sudan [101, 102]. By not testing for the degree of each variable explicitly, the tester showed potential to perform better than the testers of [20, 19, 50, 6]. Yet a key element was missing in the analysis of the tester, which was finally remedied by the work of Arora and Safra [6]. The observation that a combination of the analysis of Rubinfeld and Sudan [101] and Arora and Safra [6] yields a test which requires only a constant number of probes is made by Arora, Lund, Motwani, Sudan and Szegedy [8]. One interesting open question is: "How large must the field size be as a function of the degree d, so that the low-degree test of Theorem 3.2.5 works". This in turn reduces to the question of seeing how small the field size may be while the Arora-Safra tester still works. The analysis given in [6] shows that the field size could be $O(d^3)$, and this can be improved to show $O(d^2)$ suffices. It seems possible that this number could be reduced to being $O(d)$.

Testing approximate polynomials. A large number of analytic functions can be closely approximated by polynomials. Moreover when computing functions over the reals one might be willing to tolerate a small amount of error in the answers. In order to make testers which apply to such situations, one requires testers which will test if a function is closely approximated by a (multivariate) polynomial. As a starting point for this one would need testers which don't depend on the nice properties of finite fields. Such a test does exist and is described by Rubinfeld and Sudan [101]. But even for univariate functions that are closely approximated by polynomials, no tester seems to be known. We feel that simple extensions of the tests given in this chapter should yield a test for approximate polynomials.

Addendum. The main open question raised above – i.e., does the low-degree test of Section 3.2.3 work over fields whose size is only linear in d – has been answered recently in the affirmative by Polishchuk and Spielman [97]. Friedl and Sudan [55] also consider the low-degree testing problem and improve the results there in two ways: They show that the characterization of polynomials over prime fields given in Lemma 3.2.1 works for every prime $p \geq d + 2$. They also give an explicit bound on the constant δ_0 in Theorem 3.2.5, showing that it works for any $\delta_0 < 1/8$, provided the field F is sufficiently large. Both the above mentioned results translate into efficient constructions of "proof systems" which are the focus of the next chapter.

4. Transparent proofs and the class PCP

The problems considered towards the end of the previous chapter were raised in the context of an interactive proof setting. We were considering the task of proving certain statements about polynomials to a verifier by writing down (or providing an oracle) some information for the verifier. The proofs so obtained were checkable very efficiently by a probabilistic verifier. In this chapter we set up this notion more formally, outlining the parameters of interest, and by exploring such *probabilistically checkable proofs* for more general statements. A particular feature of interest will be the number of bits of the proof that are examined by the probabilistic verifier. A second parameter of interest is the number of random bits used by the verifier to verify the proof.

We show how the results of the previous chapter can be translated to get probabilistically checkable proofs of fairly general statements – namely, statements of the form $x \in L$ where L is a language in NP. The translation uses the work of Babai, Fortnow and Lund [20] and Babai, Fortnow, Levin and Szegedy [19] which shows that the problem of testing satisfiability of construction rules in NP-complete for a certain choice of parameters. The probabilistically checkable proofs that result can be verified by a probabilistic verifier who tosses $O(\log n)$ coins and probes the proof in $O(\text{polylog} \, n)$ bits.

Next we outline the idea of "recursive" proof checking of Arora and Safra [6]. The idea shows that if the proof systems are restricted to obey a certain format, then they can be composed as follows: If a proof system examines $q_1(n)$ bits of a proof, and another one examines $q_2(n)$ bits of a proof, then they can be composed to get a proof system which examines $q_2(q_1(n))$ bits of the proof. Furthermore the amount of randomness used by the composed system grows as $r_1(n) + r_2(q_1(n))$ where $r_1(n)$ and $r_2(n)$ are the number of random bits used by the two proof systems.

The proof system obtained from the results of the previous chapter can be shown to conform to the restrictions and this gives us one place to start from. We also develop a second proof system which examines only a constant number of bits in the proof but uses many random bits. The composition idea shows how to compose these proof systems with each other, eventually giving proof systems where the verifier tosses $O(\log n)$ coins and looks at only constantly many bits in the proof to verify it. The results of this chapter are due to Arora, Lund, Motwani, Sudan and Szegedy [8].

4.1 Definitions

The basic task of this chapter is the construction of proof systems which "magnify errors". Such proof systems should have the feature that if a statement is true then the proof system should admit error-free proofs of the statement. On the other hand, any "proof" of an incorrect statement should be riddled with errors. Formalizing this notion takes some effort and here we present two efforts which make this notion precise.

The first notion we study is that of "transparent proofs" due to Babai, Fortnow, Levin and Szegedy [19]. Babai et al., [19], achieve this formalism by restricting the running time of probabilistic verifier. Such a restriction implies that evidence of the fallacy of a statement must be scattered densely in any proof of an incorrect statement (since in very little time, the verifier is able to find evidence of the mistake in the proof). Yet, when we consider statements of the type $x \in L$, a verifier that attempts to verify this statement needs to be given at least enough time to read x. Thus it seems that the running time of the verifier would need to be at least linear in the input size. Babai et al. get around this by expecting the "theorem" to be also presented in a "transparent" form i.e., they expect the input x to be presented in an error-correcting encoding. The following definition is presented somewhat informally.

Definition 4.1.1 (transparent proof: [19]). *A pair of strings (X, π) with X being a "theorem-candidate" and π a "proof-candidate" is said to be in transparent form if X is encoded in an error-correcting code and the pair (X, π) can be verified by a probabilistic verifier in time polylogarithmic in the size of the theorem plus proof and the verifier is given random access to the strings X and π. In particular, if X is the encoding of a correct theorem, there must exist a proof π which will be accepted by the verifier for all its random choices and if X is close to the encoding of a false theorem, or not close to the encoding of any valid statement, then it must be rejected by the probabilistic verifier with high probability.*

Babai, Fortnow, Levin and Szegedy [19], based on the work of Babai, Fortnow and Lund [20], show that all theorems and proofs can be placed in a transparent form by increasing their size by a slightly superlinear factor. One interesting aspect of this formalism is the rather "blind" nature of the verification process. The verifier at the end of its verification, has little idea of what the statement being proved is, and what the proof looks like. The only guarantee it is able to give is that the two are consistent with each other. This rather surprising nature of these proof systems will turn out to be useful in Section 4.3.

The next notion we study is that of *probabilistically checkable proofs* due to Arora and Safra [6]. Instead of characterizing the "transparency" of a proof system using the running time of a verifier, they characterize it using

the number of bits of a proof that are examined by the verifier to verify the
proof, or its query complexity. This parameter was first highlighted by the
work of Feige, Goldwasser, Lovasz, Safra and Szegedy [50]. By examining
this parameter, and allowing the running time of the verifier to be fairly
large, they do not need inputs to be presented in an error-correcting code.
Moreover, any verifier for NP statements would need to have the ability to
address the entire proof and this places an inherent logarithmic lower bound
on the running time of the verifier. The number of bits of the proof examined
by a verifier has no such inherent lower bounds and thus allows for a much
more sensitive characterization of the quality of proof systems. A second
parameter, also highlighted by Feige et al, [50], examined by Arora and Safra
is the number of random bits used by the verifier. This gives an implicit bound
on the size of the proof and is also motivated by some of the applications of
such proofs systems (see the work of Feige, Goldwasser, Lovasz, Safra and
Szegedy [50] (see also Chapter 5). We define the notion in terms of languages
which have efficient probabilistically checkable proofs.

Definition 4.1.2 (PCP: [6]). *A language L is in the class $PCP(r(n), q(n))$
if there exists a tester T such that $\forall x \in \{0, 1\}^n$, we have:*

- *If $x \in L$, then there exists a proof π such that $T(r, x, \pi)$ outputs PASS, for
  all $r \in \{0, 1\}^{O(r(n))}$.*
- *If $x \notin L$ then for all proofs π', $T(r, x, \pi')$ outputs FAIL for at least half the
  strings $r \in \{0, 1\}^{O(r(n))}$.*

*Furthermore, for any fixed value of r, $T(r, x, \pi)$ depends on only $O(q(n))$ bits
of π, and its running time is bounded by $\text{poly}(n, r(n), q(n))$.*

The result of Babai et al. [19], improving on [20], can be viewed in this
setting as showing NP $\subset$ PCP(polylog n, polylog n). (The implicit guarantee
on the proof size obtained from this characterization is weaker than that
shown by Babai et al. The proof sizes as obtained by Babai et al. are nearly
linear in the size of any witness of $x \in L$.) Feige et al. [50], improved on [20]
differently to show that NP $\subset$ PCP($\log n \log\log n$, $\log n \log\log n$), but their
proofs are superpolynomial in size. Arora and Safra were the first to bring
both parameters below the "logarithmic" level, thus allowing for an exact
characterization: they show NP = PCP($\log n$, polyloglog n). In this chapter
we work towards showing NP = PCP($\log n$, $O(1)$).

From here onwards we will use the words "transparent proofs" and "probabilistically checkable proofs" interchangeably. The notion we will use will be
that of Arora and Safra [6] and precise statements will always be made in
terms of PCP.

4.2 A transparent proof for languages in NP

The first probabilistically checkable proof we will describe follows in a very simple manner from the following characterization of NP in terms of polynomial construction sequences of very short width and length (both are logarithmic in the length of the input). The characterization is implicit in the work of [20, 19].

Lemma 4.2.1 ([20, 19]). *Given a language $L \in NP$ and an instance $x \in \{0,1\}^n$, a construction rule of length $\log n$ and width $\log n$ for degree $\log^2 n$ polynomials in $\Theta(\frac{\log n}{\log\log n})$ variables from a finite field F of size $O(\log^2 n)$, can be computed in polynomial time, with the property that the construction rule is satisfiable if and only if ϕ is satisfiable.*

The proof of this lemma is included in the appendix. This gives us the first transparent proof of NP as follows:

Lemma 4.2.2. $NP \subseteq PCP(\log n, \text{polylog}\, n)$

Proof: By Lemma 4.2.1 we know that given a language $L \in$ NP and an input x of length n, we can compute in polynomial time a construction rule $r_1, \ldots, r_l$ of width $O(\log n)$ which is satisfiable if and only if $x \in L$. By Theorem 3.4.2, we can construct proofs of satisfiability of $r_1, \ldots, r_l$ which has size $\text{poly}(n)$, where the tester uses $O(\log n)$ bits of randomness, and probes the proof in $O(\text{polylog}\, n)$ bits.   □

4.3 Recursive proof checking

The proof system developed in Lemma 4.2.2 has nicer properties than just verifiability using $O(\text{polylog}\, n)$ bits. One particular feature of the proof system is its ability to perform almost "blind checks" i.e., the proof system could have been modified so that it is presented with a pair of inputs (X, π) and by making very few probes into X and π the verifier could have established the consistency of π as proof $x \in L$ where X encodes x in an error-correcting code. A second feature that comes out of the work put into Theorem 3.4.2 is the following: If proof is written as an array indexed by the questions asked by the tester and whose contents reflect the answers of the oracle to the questions, then the proof can be verified by looking at a *constant* number of entries of the array, where each entry is $O(\text{polylog}\, n)$ bits long. We call this the property of having "segmented" proofs.

The latter property implies that the verifier, in order to verify $x \in L$ tosses $O(\log n)$ coins and then its task reduces to verifying that a constant number of entries that it reads $y_1, \ldots, y_c$ from the proof table satisfy some simple computation performed by it (i.e., $T(y_1, \ldots, y_c)$ outputs PASS). The key idea behind the notion of recursive proof checking is to use the "blind checkability" of such proof systems to obtain a proof that $T(y_1, \ldots, y_c)$ outputs PASS without reading $y_1, \ldots, y_c$. Since the strings y_i are of length $O(\text{polylog } n)$, a proof of such a fact would (hopefully) be a table of size $\text{poly}(\text{polylog } n)$ whose entries are $O(\text{polyloglog } n)$ bits long. The recursive testing of this fact would thus hopefully probe a constant number of entries in these small tables, giving proofs verifiable with $O(\text{polyloglog } n)$ probes.

One problem with the immediate implementation of this idea is that the guarantee on blind checkability assumes that the input x for a statement of the type $x \in L$ is presented in an error-correcting encoding. But the recursion involves statements of the type $y_1 \cdot y_2 \cdots y_c \in L$, where the prover can only provide individual encodings of the y_i's. It turns out though that the proof verification system of Babai, Fortnow, Levin and Szegedy [19] can be modified to get the "blind checkability" even when the input is given in the form of a constant number of encoded entries and this was first observed by Arora and Safra [6].

Thus we restrict our attention to proof systems which have both the properties considered above, namely, segmented proofs and blind checkability when the input is presented by a constant number of encoded pieces and show how to compose such proofs systems to achieve proof systems with improved query complexity.

We formalize the notion of tables and encodings next. Some care must be taken to define the behavior of the proof systems in the context of working with encoded inputs. In particular, one would need to handle the case where the supposed encoding of the input is not really close to any valid encoding. One would like the proof system to reject such a proof and this notion is made precise via the notion of the inverse mapping of an encoding scheme.

Definition 4.3.1 (segmented tables). *An $s(n) \times q(n)$-table τ is a function from $[s(n)]$ to $\{0,1\}^{q(n)}$. The values $\tau(i)$, $1 \leq i \leq s(n)$ will be referred to as the segments of τ.*

Definition 4.3.2 (Encoding/Decoding). *An $s(n) \times q(n)$-encoding scheme E encodes n bit strings into an $s(n) \times q(n)$-table. A decoder E^{-1} for E is a function which takes tables of size $s(n) \times q(n)$ and produces n bit strings, such that $E^{-1}(E(x)) = x$. Notice that in general most elements of the domain of E^{-1} are not constructed from applying E to any string, yet E^{-1} maps them to strings from the domain of E. This is supposed to resemble the task of performing error-correction and then decoding.*

Definition 4.3.3 (restricted PCP). *For functions $r, q : \mathcal{Z}^+ \to \mathcal{Z}^+$, a language L is in the class $rPCP(r(n), q(n))$ if there exists a constant c_L such that $\forall$ integers c, $\exists s : \mathcal{Z}^+ \to \mathcal{Z}^+$ satisfying $s(n) \leq 2^{r(n)}$, and an $s(n) \times q(n)$-encoding scheme E with a decoder E^{-1} and a tester T, such that given c n-bit strings $x_1, \ldots, x_c$, the following are true:*

- If $x_1 \cdot x_2 \cdots x_c$ *(the concatenation of x_1 through x_c) is contained in L, then there exists a $s(n) \times q(n)$-proof table π such that for all random choices of $r \in \{0,1\}^{r(n)}$, the tester T_r accepts $E(x_1), \ldots, E(x_c)$ and π.*
- If $x_1 \cdot x_2 \cdots x_c \notin L$, then for all proofs π' and for all tables $\tau_1, \ldots, \tau_c$ such that $E^{-1}(\tau_i) = x_i$, T_r rejects $\tau_1, \ldots, \tau_c, \pi'$ for at least half the choices of $r \in \{0,1\}r(n)$.*

Moreover, the output of the tester T for a fixed choice of r depends on only $c \cdot c_L$ segments of the input tables and the proof table, and can be computed by a circuit whose size is $poly(q(n))$. Lastly, E should be polynomial time computable. (Notice that E^{-1} need not be computable efficiently.)

The following lemma is based directly on the work of Arora and Safra [6] and shows that two rPCP proof systems can be composed to get potentially more efficient rPCP proof systems.

Lemma 4.3.1. *If $NP \subset rPCP(r_1(n), q_1(n))$ and $NP \subset rPCP(r_2(n), q_2(n))$ then $NP \subset rPCP(r(n), q(n))$ where $r(n) = r_1(n) + r_2(q_1(n)^{O(1)})$ and $q(n) = q_2(q_1(n)^{O(1)})$.*

The proof of this lemma is straightforward given the definition of rPCP. This proof is deferred to the appendix.

4.4 Restricted PCP's for languages in NP

The characterization of NP in terms of polynomial sequences can be strengthened so as to be able to use encoded inputs. The encoding we will choose for the inputs will be the (m, h)-polynomial extension encoding. Recall that for any choice of h^m values $\{v_z\}_{z \in H^m}$, there exists a polynomial $g : F^m \to F$ of degree at most mh such that $g(z) = v_z$ for $z \in H^m$.

Definition 4.4.1. *For an n bit string x where $n = h^m$, the encoding $E_{m,h,F}$ encodes x according to the (m, h) polynomial extension encoding (i.e., finds a polynomial g which agrees with x on the space H^m and writes out its value over all the points in F^m). The inversion scheme $E^{-1}_{m,h,F}$ we will pick for E will map every function from F^m to F, to the closest degree mh polynomial (ties may be broken arbitrarily) and use its values over the domain H^m as the value of the inverse.*

Theorem 4.4.1. *Given a language $L \in NP$, a constant c and an input length n, a polynomial construction rule for degree $\log^2 n$ polynomials $g^{(0)}, \ldots, g^{(l)}$ $(g^{(i)} : F^{m+1} \to F)$ of length $O(\frac{\log n}{\log \log n})$ and width $O(\log n)$ can be constructed in polynomial time such that: $\exists g^{(0)} s.t. g^{(l)} \equiv 0$ if and only if $x_1 \cdots x_c \in L$ where $x_i = E^{-1}_{m,h,F}(g^{(0)}|_{z_1=i})$ (where the notation $g^{(0)}|_{z_1=i}$ represents the polynomial on m variables obtained by setting the value of the first variable z_1 to a value $i \in F$). Lastly $m = \Theta(\frac{\log n}{\log \log n})$, $|F| = polylog n$ and $h = \log n$.*

The proof of this statement is included in the appendix. This allows us to construct our first restricted PCP proof system.

Lemma 4.4.1. $NP \subset rPCP(\log n, \text{polylog } n)$.

Proof: The rPCP proof π will consist of $g^{(0)}$, $g^{(l)}$ and an oracle O according to Theorem 3.4.2 which allows us to verify using a constant number of probes into O, $g^{(l)}$ and $g^{(0)}$ that the construction rules have been obeyed. The length of the longest segments in this proof are the entries of O which are of length $O(\text{polylog } n)$ bits long. The tester T is essentially the same as the tester of Theorem 3.4.2 who verifies that $g^{(l)}$ has been obtained from $g^{(0)}$ by following the construction rules. In addition the tester will ensure that $g^{(l)} \equiv 0$. The number of random bits is essentially the same as in the tester of Theorem 3.4.2, which is $O(m \log |F| + l) = O(\log n)$. $\qquad \square$

Notice that by composing this system with itself, using Lemma 4.3.1, we can obtain NP $\subset$ rPCP($logn$, polyloglog n) and by continuing the process, we can get NP $\subset \log^{(c)} n$, for any constant c (where $\log^{(c)}$ denotes the cth iterated logarithm function). Yet this does not seem to suffice to show a result of the form NP $\subset$ rPCP($\log n$, $O(1)$). In order to show such a result we need *some* protocol where the number of bits read is independent of the length of the statement being proved. In the next section we describe such a proof system.

4.5 A long and robust proof system

In this section, we construct long but highly transparent proofs of membership for languages in NP. The essential idea behind reducing the size of the table entries is the use of very low degree polynomials. In fact, all the results of this section are derived from polynomials of degree one. This results in the need to use many variables in the polynomials, so as to encode sufficient amounts of information. This in turn, is what causes the explosion in the amount of randomness by exponential factors.

4.5.1 Preliminaries: linear functions

We first review some of the key facts about linear functions. Some of the facts mentioned here might follow from the work done on higher degree polynomials in Chapters 3 and 2, but we mention them here anyway to reemphasize the basic properties that will be used in the rest of this section.

Definition 4.5.1 (linearity). *A function $A : F^m \to F$ is called* linear *if there exist $a_1, \ldots, a_m \in F$ such that $A(x_1, \ldots, x_m) = \sum_{i=1}^m a_i * x_i$.*

The following is a well-known fact.

Fact 4.5.1. A function $A : F^m \to F$ is linear if and only if for all $x, y \in F^m$, $A(x + y) = A(x) + A(y)$.

The fact above was strengthened very significantly by Blum, Luby and Rubinfeld [36] who show that the property used above is a very "robust" one, and can hence be used to construct testers for the family of linear functions.

Lemma 4.5.1 (linearity tester: [36]). *If* $\tilde{(A)} : F^m \to F$ *satisfies*

$$\Pr_{x, y \in_U F^m} \left[\tilde{A}(x + y) = \tilde{A}(x) + \tilde{A}(y) \right] \geq 1 - \delta/2$$

then $\exists$ *a linear function* A *such that* $d(A, \tilde{A}) \leq \delta$, *provided* $\delta \leq 1/3$.

Blum, Luby and Rubinfeld [36] also show that the family of linear functions is self-correctable. In fact, they show that the value of a linear function can be computed correctly anywhere, using *two* calls to a function that is close to it.

Lemma 4.5.2 (linear self-corrector: [36]). *If* $\tilde{A}$ *is* δ-*close to a linear function* A, *then for all* $x \in F^m$

$$\Pr_{y \in_U F^m} \left[A(x) = \tilde{A}(y + x) - \tilde{A}(y) \right] \geq 1 - 2\delta$$

The important point about the lemmas above is that both hold for all finite fields and in particular $GF(2)$. This immediately allows us to create error-correcting codes with very interesting error detection and correction properties. The encoding of n-bits $a_1, \ldots, a_n$ is the 2^n bit string $\{A(x)\}_{x \in Z_2^n}$. The linearity tester becomes a randomized error detector and the self-corrector becomes a randomized error correcting scheme. These properties will now be used in the next section to construct proofs of satisfiability.

4.5.2 Long proofs of satisfiability

In this section we consider a 3-CNF formula ϕ on n variables $v_1, \ldots, v_n$ and m clauses $C_1, \ldots, C_m$. The prover is expected to prove the satisfiability of ϕ by providing a satisfying assignment $a_1, \ldots, a_n$, encoded in a suitable error-correcting code.

The coding scheme we choose here is based on the scheme of coding via linear functions, that we touched upon in the previous section. We first develop some notation. The assignment $a_1, \ldots, a_n$ will be denoted by the vector $a \in Z_2^n$.

Definition 4.5.2. *For vectors* $x \in Z_2^l$ *and* $y \in Z_2^m$, *let* $x \circ y$ *denote the* outer product $z \in Z_2^{lm}$, *given by* $z_{ij} = x_i * y_j$. *Note that although* z *is an* $l \times m$ matrix, *we will sometimes view it as an* lm-dimensional vector. *The exact view should be clear from the context.*

Let $b = a \circ a$ and let $c = a \circ b$. Further let $A : \mathcal{Z}_2^n \to \mathcal{Z}_2$, $B : \mathcal{Z}_2^{n^2} \to \mathcal{Z}_2$ and $C : \mathcal{Z}_2^{n^3} \to \mathcal{Z}_2$ be the linear functions whose coefficients are given by a, b and c.

$$A(x) = \sum_{i=1}^{n} a_i * x_i$$

$$B(y) = \sum_{i=1}^{n} \sum_{j=1}^{n} b_{ij} * y_{ij}$$

$$C(z) = \sum_{i=1}^{n} \sum_{j=1}^{n} \sum_{k=1}^{n} c_{ijk} * z_{ijk}$$

The encoding scheme for a that we choose is the following: The prover writes down the values of the functions A, B and C explicitly for each input. The intuition for choosing this encoding is the following:

1. By using the results on linearity testing it should be possible to verify the authenticity of such codewords.
2. Given the information specified above correctly, one can compute the value of *any* degree 3 polynomial in n variables at the point a.
3. A 3-CNF formula should be closely related to degree 3 polynomials.

We now provide precise statements of the claims above and prove them.

Lemma 4.5.3. *Given functions $\tilde{A}$, $\tilde{B}$ and $\tilde{C}$, and a constant $\delta > 0$ there exists a tester T, and a constant c such that:*

- *If there exists a vector a such that $\tilde{A}$, $\tilde{B}$ and $\tilde{C}$ give the encoding of a, then $T(r, \tilde{A}, \tilde{B}, \tilde{C})$ outputs PASS for all $r \in \{0,1\}^{O(n^3)}$.*
- *If for all vectors a, at least one of the distances $d(A, \tilde{A})$, $d(B, \tilde{B})$ and $d(C, \tilde{C})$ is not bounded by δ, then $T(r, \tilde{A}, \tilde{B}, \tilde{C})$ outputs FAIL, for at least half the random strings $r \in \{0,1\}^{O(n^3)}$.*

Furthermore, for any fixed choice of r, T's output depends on at most c values of $\tilde{A}$, $\tilde{B}$ and $\tilde{C}$.

Proof: The tester T first tests that the functions $\tilde{A}$, $\tilde{B}$, $\tilde{C}$ are linear functions, using the tester from Lemma 4.5.1. This yields strings a, b and c such that: if A, B and C are the linear function with coefficients a, b and c respectively, then $d(A < \tilde{A})i$, $d(B, \tilde{B})$ and $d(C, \tilde{C})$ are all bounded by δ. This tester needs to probe $\tilde{A}$, $\tilde{B}$ and $\tilde{C}$ in $O((\frac{1}{\delta})^2)$ places. Further note that at this point we could use the self-corrector of Lemma 4.5.2 we can compute the functions A, B and C at any point correctly with high probability.

At this point the only aspect left to be tested is that $b = a \circ a$ and that $c = a \circ b$. We now test that these properties hold. These tests will be based on the randomized algorithm for verifying matrix products, due to Freivalds [56]. Consider the $n \times n$ matrix X such $X_{ij} = b_{ij}$ and let Y be the $n \times n$

matrix obtained by viewing $a \circ a$ as an $n \times n$ matrix. The property we wish to verify is that $X = Y$. The idea of Freivalds' matrix multiplication checker is to consider a random vector $x \in \mathcal{Z}_2^n$ and verifying that $x^T X = x^T Y$. It can be shown that if $X \neq Y$ then this products differ with probability at least half.

Further, consider a randomly chosen vector $y \in \mathcal{Z}_2^n$ and the products $x^T X y$ and $x^T Y y$. If $x^T X \neq x^T Y$ then these products differ with probability half. Thus with probability at least a quarter, we have that $x^T X y \neq x^T Y y$, if $X \neq Y$. But now consider the product $x^T X y$: this is equal to $B(x \circ y)$, and the product $x^T (a \circ a) y$ equals $A(x) * A(y)$. Thus the identity can be tested by evaluating the functions A and B at three points in all. The process can be repeated constantly many times to get high enough confidence. A similar test checking that $C(x \circ y) = A(x) * B(y)$ concludes the test.   □

Next consider the task of evaluating any degree 3 polynomial f at the point $a_1, \ldots, a_n$. f can be written as

$$
\begin{aligned}
f(a_1, \ldots, a_n) &= \alpha + \sum_{i \in S_1} a_i + \sum_{(i,j) \in S_2} a_i * a_j + \sum_{(i,j,k) \in S_3} a_i * a_j * a_k \\
&= \alpha + \sum_{i \in S_1} a_i + \sum_{(i,j) \in S_2} b_{ij} + \sum_{(i,j,k) \in S_3} c_{ijk} \\
&= \alpha + A(\mathcal{X}(S_1)) + B(\mathcal{X}(S_2)) + C(\mathcal{X}(S_3))
\end{aligned}
$$

(where S_1, S_2 and S_3 are sets that depend only of f and $\mathcal{X}(S_1)$, $\mathcal{X}(S_2)$ and $\mathcal{X}(S_3)$ are the characteristic vectors of these sets). Thus any degree 3 polynomial can be evaluated at a by computing A. B and C at one point each. Next we show that 3-CNF formulae are closely related to degree 3 polynomials.

Lemma 4.5.4. *Given a 3-CNF formula ϕ, and an assignment $a_1, \ldots, a_n$, a degree 3 polynomial $\tilde{\phi} : \mathcal{Z}_2^n \to zt$ can be constructed (without knowledge of the assignment) such that*

– If $a_1, \ldots, a_n$ satisfies ϕ, then $\tilde{\phi}(a_1, \ldots, a_n) = 0$.
– If $a_1, \ldots, a_n$ does not satisfy ϕ then $\tilde{\phi}(a_1, \ldots, a_n) = 1$ with probability $1/2$.

Proof: We first arithmetize every clause C_j into an arithmetic expression $\tilde{C}_j$ over $\mathcal{Z}_2$ (over the same set of variables), so that C_j is satisfied by a if and only if $\tilde{C}_j$ evaluates to zero. This is done as follows: If C_j is the clause $v_1 \vee v_2 \vee \neg v_3$ then $\tilde{C}_j$ will be the expression $(1 - v_1) * (1 - v_2) * v_3$. Notice that each clause gets converted in this fashion to a degree 3 polynomial and the whole formula ϕ is satisfied only if each expression $\tilde{C}_j$ evaluates to zero at $v_i = a_i$.

Now consider taking the inner product of the vector $< \tilde{C}_1, \ldots, \tilde{C}_m >$ with a randomly chosen m-bit vector r. If the vector $< \tilde{C}_1, \ldots, \tilde{C}_m >$ is not identically zero then the inner product will be non-zero with probability half. Thus if we let $\tilde{\phi}$ be the inner product i.e., $\sum_{j=1}^m r_j * \tilde{C}_j$ then $\tilde{\phi}$ is a degree three polynomial in the n variables which satisfies the conditions required by the Lemma. $\square$

Thus we are in a position to prove the following lemma.

Lemma 4.5.5. $NP \subseteq PCP(\mathrm{poly}(n), 1)$.

Proof: For any language L and fixed input length n, we create a 3-CNF formula ϕ such that

$$\forall w \in \{0,1\}^n \exists y \text{ such that } \phi(w, y) \text{ is true } \Leftrightarrow w \in L$$

We then expect the prover to encode the string $a = w \cdot y$ using the encoding mechanism (i.e., the functions A, B and C) as constructed in this section. The tester T first verifies that the encoding describes a valid assignment, using Lemma 4.5.3 and then verifies that it corresponds to a satisfying assignment of ϕ by creating $\tilde{\phi}$ as described in Lemma 4.5.4. Notice further that the parity $a = w \cdot y$ on any subset of the bits can be expressed as the value of A at a certain point. The tester T uses this fact to verify that the initial portion of a is the same as w. The tester picks a random subset of the bits of w and compares its parity with the parity of a on the same subset of bits. If the initial portion of a is different from w then this test will detect this with probability half. This test is repeat enough times to get large enough probabilities of detecting cheating.

Thus tester T rejects the proof (i.e., the functions A, B and C), with probability half if $w \notin L$ and accepts with probability one if $w \in L$. $\square$

Lemma 4.5.6. $NP \subset rPCP(\mathrm{poly}(n), O(1))$

Proof: To convert the proof system of Lemma 4.5.5 to a $rPCP(\mathrm{poly}(n), O(1))$ proof system we observe that if the input $w = w_1 \cdot w_2 \cdots w_c$ are each encoded by the parities of all their subsets, then the last phase of the tester's verification process, just compares entries from the encodings of w_1 etc. with the value of A at some point. Thus there exists an encoding scheme E and a proof π such that the tester looks at $O(1)$ bits from the encodings of $w_1, \cdots, w_c$ and $O(1)$ bits of the proof π and verifies that $w_1 \cdot w_2 \cdots w_c \in L$. $\square$

4.6 Small proofs with constant query complexity: recursion

The results of the previous two sections can be combined using the recursion lemma, Lemma 4.3.1 to get proofs which combine the best of both the proof systems.

Theorem 4.6.1. $NP \subset rPCP(\log n, O(1))$

Proof: Using Lemmas 4.3.1 and 4.4.1, we get that

$$\text{NP} \subset \text{rPCP}(\log n, \text{polyloglog} \, n).$$

Now using this result and Lemma 4.5.6, we see that $\text{NP} \subset \text{rPCP}(\log n, O(1))$.
□

The result can be extended to get the following theorem for general NTIME classes.

Theorem 4.6.2. *If $L \in NTIME(t(n))$, then $L \in PCP(\log(t(n) + n), O(1))$.*

Thus the following corollaries become immediate.

Corollary 4.6.1. $NE = PCP(n, O(1))$

Corollary 4.6.2. $NEXPTIME = PCP(\text{poly}(n), O(1))$

4.7 Discussion

Most of the work prior to 1992 [20, 19, 50] were essentially focussed on deriving PCP containments for nondeterministic time bounded classes. While these results lie at the core of the constructions described in this chapter, none of them translate directly into any rPCP results. The only work prior to 1992, which can be described in terms of rPCP systems are the results on "constant-prover proof systems" (i.e., multiple prover proof systems which allow only one round of interaction with upto a constant number of provers) [29, 53, 37, 85, 48]. In particular the results of Lapidot and Shamir [85] and Feige and Lovasz [48] can be scaled down to get $\text{rPCP}(\text{polylog} \, n, \text{polylog} \, n)$ protocols for NP. The $\text{rPCP}(\log n, \text{polylog} \, n)$ protocol given here is somewhat inspired by the works of Lapidot and Shamir [85] and Feige and Lovasz [48] on parallelizing the MIP = NEXPTIME protocol, and in particular shares similar goals, but the protocol and proof are new to Arora, Lund, Motwani, Sudan and Szegedy [8]. In particular, the amount of randomness used in the protocols of [85, 48] seems to be superlogarithmic and [8] are able to reduce this to $O(\log n)$. Moreover, their final protocols does not seem to be able to handle the situation where the input comes in a constant number of error-corrected pieces. The $\text{rPCP}(\text{poly}(n), O(1))$ protocol discussed in this chapter is new to Arora, Lund, Motwani, Sudan and Szegedy [8]. The recursive proof construction technique described here is almost entirely due to the work of Arora and Safra [6], modulo the formalism which may be different here.

Open Questions. The most important question that does remain open is what is the smallest number of bits that need to be read from a transparent proof to achieve a fixed probability of detecting a false proof. In the next chapter a connection is pointed out between the PCP proof systems and 3SAT. This connection shows that if the probability of detecting cheating is allowed to be an arbitrarily small constant, then reading 3 bits of the proof suffices. Moreover, if the error of the verifier is expected to be one sided, then 3 bits are necessary (the computation of a verifier when it reads only 2 bits can be equated to the satisfiability of a 2-SAT formula). Lastly, in this regard, it may be pointed out that if the error of the verifier is allowed to be two-sided then even reading two bits suffices.

Our result shows that any proof can be converted into a transparent proof which is within a polynomial factor of the size of the original proof. In contrast to this, the transparent proofs of Babai, Fortnow, Levin and Szegedy [19] are nearly linear in the size of the original proof. This raises the question of whether the proofs of this section can be compressed into a nearly linear size. This question seems to get mapped down to the question of the efficiency of the low-degree test and the question about the field sizes required for the Arora-Safra Tester that is raised at the end of Chapter 3.

Addendum. In the last few years the PCP technology has seen enormous amounts of progress – we attempt to summarize this briefly here. Bellare, Goldwasser, Lund and Russell [28] initiated the attempt to reducing the constants in the proof systems described here. They show that the query complexity of a proof system can be made as small as 30 bits while guaranteeing that incorrect proofs will be rejected with probability half. This number was further reduced by Feige and Kilian [47] and by Bellare, Goldreich and Sudan [27] who report that 16 bits suffice. Much of this improvement comes about due to an improvement in the parameters associated with a constant prover proof system for NP [28, 47, 108, 98]. A second factor in the improvements has been improvements in analysis of the "linearity test" – which also has some fairly tight analysis now due to [28, 26]. A different objective has been the blowup in the size of a transparent proof as compared with the original proof. This was the original objective of Babai, Fortnow, Levin and Szegedy [19] who set the goal of getting a transparent proof with a blowup of only $O(n^{1+\epsilon})$. This goal was met by Polishchuk and Spielman [97] who give a proof system with constant query complexity and $O(n^{1+\epsilon})$ size proofs. A third direction of research combines the two objectives above trying to combine the goal of small proof sizes with small (explicit) query complexity. Friedl and Sudan [55] report proofs with such explicit parameters, giving proofs with $O(n^{2+\epsilon})$ size with a query complexity of 165 bits.

5. Hardness of approximations

The notion of NP-completeness (Cook [41], Levin [86] and Karp [80]) was developed primarily as an attempt to explain the apparent intractability of a large family of combinatorial optimization problems. The resulting theoretical framework (cf. [60]) was defined mainly in terms of decision problems obtained by imposing bounds on the value of the objective function. This permitted the development of an elegant body of results and the formalism sufficed for the purposes of classifying the complexity of finding optimal solutions to a wide variety of optimization problems.

Attempts to extend this analysis to the task of finding approximate solutions to the same set of problems, was not very successful. Problem which seemed equivalent when the goal was to find exact solutions, seems to break apart into problems of widely varying complexity when the goal was relaxed to that of finding approximate solutions. Some problems like the knapsack problem have extremely good approximation algorithms [59]. Other problems have algorithms where the error of approximation can be made arbitrarily small, but the penalties paid for improved solutions are heavy. An example of such a problem is the task of minimizing the makespan on a parallel machine - a scheduling problem studied by Hochbaum and Shmoys [73]. Yet other problems like the Euclidean TSP and vertex cover seemed approximable to some constant factor but not arbitrarily small ones; and finally we have problems which seem no easier to approximate than to solve exactly e.g. Chromatic number.

Some initial success was obtained in showing the hardness of even approximating certain problems: For the traveling salesman problem without triangle inequality Sahni and Gonzalez [103] showed that finding a solution within any constant factor of optimal is also NP-hard. Garey and Johnson [58] showed that the chromatic number a graph could not be approximated to within a factor of $2 - \epsilon$. They also show that if the clique number of a graph cannot be approximated to within some constant factor, then it cannot be approximated to within any constant factor. Hochbaum and Shmoys [71, 72] study some min-max problems where they show tight bounds on the factor to which these problems may be approximated unless NP = P.

The lack of approximation preserving reductions among optimization problems seemed to isolate these efforts and the search for such reductions became the goal of a wide body of research [10, 11, 12, 95]. The most successful of these efforts seems to be the work of Papadimitriou and Yannakakis [93] where they used a syntactic characterization of NP due to Fagin [45] to define a class called MAX SNP. They also defined a particular approximation preserving reduction called the L-reduction (for *linear* reductions) used these reductions to find complete problems for this class. All problems in this class were approximable to some degree, and the complete problems for the class seemed hard to approximate to arbitrarily small factors. The class MAX SNP seemed to provide a much need framework to deal with approximation problems and this was evidenced by the large number of problems which were subsequently shown to be hard for this class [93, 94, 32, 33, 43, 77, 31, 79].

Yet, the hardness of MAX SNP seemed like a weaker condition than hardness for NP, and except for the chromatic number no unweighted combinatorial problem could be shown to being hard to approximate to some degree. It hence came as a big surprise when Feige, Goldwasser, Lovasz Safra and Szegedy [50], were able to show hardness (under a slightly weaker assumption than $P \neq NP$) of approximating the clique number of graphs to within constant factors. The hardness result used recent results in the area of interactive proofs in a very clever but simple manner, thus serving to illustrate the power of the machinery that had been built in the area of interactive proofs. Here, by showing an equally simple connection between such results and MAX SNP, we are able to show hardness results for all MAX SNP hard problems.

In the following sections we will first define the notions of approximation problems and lay out the various goals that could be set for an approximation problem. In the following section we delve into the class MAX SNP and outline some of its features. We then go on to relate the notion of probabilistically checkable proofs with MAX SNP. We do so by formulating the task of finding a PCP as an optimization problem in MAX SNP. The gap in the definition of PCP creates a gap in the optimization problem, which yields a hardness result even for the approximate version to this problem.

5.1 Optimization problems and approximation algorithms

The following is the definition of a NP optimization problem.

Definition 5.1.1 (optimization problem). *An instance I of a NP optimization problem Π consists of the pair (S, value) where S represents the solution space and $\text{value} : S \rightarrow \Re$ is a polynomial time computable function referred to as the objective function. The goal of the problem maybe any one of the following:*

1. *Given a real number k, determine if there exists a solution s such that $\text{value}(s) \geq k$ (or $\text{value}(s) \leq k$ for minimization problems).*

2. *Find the maximum (minimum) achievable value of* value *over S. This quantity is denoted OPT(I).*

3. *Find the solution $s \in S$ which maximizes (minimizes)* value(s). *Typically, the solution space S is of the form $\{0,1\}^n$ and the function* value *has a description length which is polynomial in n.*

It turns out that for many interesting problems (and in particular, the NP-complete ones), the above three goals are equivalent under polynomial time reductions. For approximation versions of the above questions, though, the problems may not remain equivalent any more. In this chapter we will use the following notion of an approximate solution for an optimization problem.

Definition 5.1.2 (approximation algorithm). *An ϵ-approximation algorithm for an NP optimization problem Π, takes an instance I as input and outputs an estimate E which satisfies*

$$\frac{E}{1+\epsilon} \leq OPT \leq (1+\epsilon)E$$

Notice that the above definition corresponds to the second of the three possible definitions of exact optimization problems. Note that in many applications, it would be more useful to produce an algorithm actually outputs a solution which comes close to the maximum value. But, since we are trying to prove negative results about the existence of such algorithms, proving it for the weaker notion is a stronger result.

Definition 5.1.3 (polynomial time approximation scheme: PTAS). *For an optimization problem Π, a polynomial time approximation scheme, takes a parameter ϵ and produces an ϵ-approximation algorithm A_ϵ for the problem Π. The running time of A_ϵ on inputs of length n is bounded by a polynomial in n. (The input here is a description of the solution space S and the function f.)*

The research efforts of the past two decades [10, 11, 12, 60, 59, 95] have broadly aimed at classifying approximation versions of optimization problems into one of the following classes:

1. Fully polynomial time approximable problems: These are problems Π, for which there exists an algorithm A, such that A takes as input an instance I of Π and an approximation factor ϵ and produces as output an ϵ-approximate estimate. The running time of A is polynomial in $|I|$ and $\frac{1}{\epsilon}$.

2. Problems with polynomial time approximation schemes.

3. Approximable problems: These are problems for which some constant ϵ exists, such that an ϵ-approximate estimate can be found in polynomial time.

4. Hard problems: These are problems for which no constant factor approximation is known.

5.2 MAX SNP: constraint satisfaction problems

The class MAX SNP was defined by Papadimitriou and Yannakakis [93] based on the syntactic definition of NP of Fagin [45] and on subsequent definition of strict-NP due to Kolaitis and Vardi [84]. The formal definitions are presented below.

Definition 5.2.1 (NP: [45]). *A predicate Π on structures I, is in NP if it can be expressed in the form $\exists S\phi(I, S)$, where S is a structure and ϕ is a first order predicate.*

(In the above definition Π is equivalent to the problem and I the instance of the problem.)

Definition 5.2.2 (SNP: [84]). *A predicate Π on structures I, is in SNP if it can be expressed in the form $\exists S\forall \overline{x}\phi(\overline{x}, I, S)$, where S is a structure and ϕ is a quantifier free predicate.*

Definition 5.2.3 (MAX SNP: [93]). *An optimization problem Π on structures I is in MAX SNP if its objective function can be expressed as*

$$\max_{S} |\{\overline{x} : \phi(\overline{x}, I, S)\}|$$

The following problem provides an alternate view of MAX SNP. It defines a combinatorial problem which turns out to be the "universal" MAX SNP problem. The combinatorial nature of the problem statement might make it an easier definition to use.

Definition 5.2.4. *A constraint of arity c is function from c boolean variables to the range $\{0, 1\}$. The constraint is said to be **satisfied** by an instantiation of its inputs if the boolean function evaluates to 1 at the instantiation.*

Definition 5.2.5 (constraint satisfaction problem). *For a constant c, an instance I of c-CSP consists of a set of constraints $C_1, \cdots, C_m$ of arity c on variables $x_1, \ldots, x_n$ where the objective function is*

$$\max_{\text{assignments to } x_1, \ldots, x_n} |\{C_i | C_i \text{ is satisfied by the assignment }\}|$$

The c-CSP is a universal MAX SNP problem in the sense that a problem lies in MAX SNP if and only if there exists a c such that it can be expressed as c-CSP. The proof of this claim is straightforward and omitted.

Papadimitriou and Yannakakis also introduced the notion of a linear reduction (L-reduction) which is an approximation preserving reduction. The notion of L-reductions allows them to find complete problems for the class MAX SNP.

Definition 5.2.6 (linear reduction: [93]). *An optimization problem Π is said to L-reduce to a problem Π' if there exist polynomial time computable functions f, g and constant $\alpha, \beta \geq 0$ such that*

1. f *reduces an instance I of Π to an instance I' of Π' with the property that $OPT(I) \leq \alpha OPT(I')$.*
2. g *maps solutions s' of I' to solutions s of I such that $|\text{value}'(s') - OPT(I')| \leq \beta|\text{value}(s) - OPT(I)|$.*

It is clear from the above definition that if there is a polynomial time algorithm for Π' with worst-case error ϵ, then there is a polynomial time algorithm for Π with worst-case error $\alpha\beta\epsilon$. Using the above definition of L-reductions, Papadimitriou and Yannakakis showed that the following problems were complete for MAX SNP. This means, in particular, that if any of the following problems has a PTAS then all problems in MAX SNP have a PTAS.

MAX 3 SAT: Given a 3-CNF formula, find an assignment which maximizes the number of satisfied clauses.

MAX 2 SAT: Given a 2-CNF formula, find an assignment which maximizes the number of satisfied clauses.

INDEPENDENT SET-B: Given a graph G with maximum degree of any vertex being bounded by a constant B, find the largest independent set in the graph.

VERTEX COVER-B: Given a graph G with maximum degree of any vertex being bounded by a constant B, find the smallest set of vertices which covers all the edges in the graph. The version of this problem with no bounds on the degree of a vertex, VERTEX COVER, is hence also hard for MAX SNP.

MAXCUT: Given a graph G find a partition of the vertices which maximizes the number of edges crossing the cut.

Further, they show that every problem in MAX SNP is approximable to some constant factor.

Lemma 5.2.1 ([93]). *For every problem Π in MAX SNP, there exists a constant ϵ such that there exists an ϵ-approximation algorithm for Π which runs in polynomial time.*

Proof: We use the universality of the c-CSP. Consider a constraint optimization problem with constraints C_1 to C_m where any constraint is a function of at most c variables. Let m' be the number of constraints which are individually satisfiable i.e., constraints for which there exists an instantiation which will satisfy them. Then $\frac{m'}{2^c} \leq OPT(I) \leq m'$. Thus an algorithm that computes m' and outputs it is a 2^c-approximation algorithm. $\qquad\square$

Even more interesting than problems in MAX SNP are the wide variety of problems that are known to be hard for this class. We compile here a list of few of them.

TSP(1,2): ([94]) Given a complete graph on n vertices with lengths on its edges, such that all edge lengths are either one or two, find the length of the shortest tour which visits all vertices at least once.

STEINER TREE$(1,2)$: ([32]) Given a complete graph on n vertices with weights on its edges, such that all edge weights are either 1 or 2, and a subset S of the vertices, find the minimum weight subgraph which connects the vertices of S.

SHORTEST SUPERSTRING: ([33]) Given a set of strings $S_1, \ldots, S_k$, over the alphabet $\{0, 1\}$, find the length of the shortest string S which contains all the given strings as substrings.

MAX CLIQUE: ([31, 50]) [1] Given a graph G find the largest clique in the graph. (This problem was shown to be very hard for MAX SNP, in that if the clique size could be approximated to within n^δ for any $\delta > 0$, then there exists a PTAS for MAX 3 SAT.)

LONGEST PATH: ([94, 79]) Given a graph G, approximate the length of the longest path in the graph to within *any* constant factor.

5.3 Non-existence of PTAS for MAX SNP hard problems

We now establish the hardness of approximating MAX SNP hard problems to within arbitrarily small factors of approximation. The results of this section are from the paper by Arora, Lund, Motwani, Sudan and Szegedy [8].

Consider a language $L \in$ NP and a transparent proof of membership of an input instance x in the language L. The question of deciding whether such a proof exists can be converted into an optimization problem as follows:

- The space of solutions will be all possible $s(n)$ bit strings, each one representing a potential proof. Each bit of the proof will be treated as an independent variable. This creates $s(n)$ variables denoted π_i.
- For each possible random string r tossed by the tester of the transparent proof, we set up a constraint $T_{r,x}$ to simulate the tester's action on the chosen random string. The constraint $T_{r,x}$ is a specification on some $O(1)$ variables from the set $\{\pi_1, \ldots, \pi_{s(n)}\}$.
- The optimization problem which questions the existence of a valid proof is:

MAX PCP:

Given $x \in \{0,1\}^n$ find $\max_{\pi \in \{0,1\}^{s(n)}} \{ |\{r| \text{ constraint } T_{r,x} \text{ is satisfied by } \pi\}| \}$

Claim. MAX PCP $\in$ MAX SNP.

---

[1] Berman and Schnitger [31] showed the hardness result mentioned here under the assumption that MAX 3 SAT did not have *randomized* PTAS. The assumption could be made weaker using some of the known derandomization techniques (say, using the idea of recycling random bits [38, 74]). The result could also be observed from the reduction of Feige, Goldwasser, Lovasz, Safra and Szegedy [50] as a special case.

Proof: By Theorem 4.6.1 we have NP = PCP($\log n, O(1)$). Thus the number of different random strings is polynomial in the input size. Hence the number of constraints is polynomial in $|x|$. Further, since each $T_{r,x}$ is a constraint on a constant number of the variables π_i, this fits the definition of a constraint satisfaction problem and thus is a MAX SNP problem. □

Claim. Approximating MAX PCP to within 10% is NP-hard.

Proof: Consider an arbitrary language $L \in$ NP and an instance $x \in \{0,1\}^n$. The MAX PCP problem deciding whether x has a transparent proof of membership will have an optimum value of either $2^{r(n)}$ if $x \in L$ or at most $2^{r(n)-1}$ if $x \notin L$. Thus a 10% approximation to the optimum value will give an answer of at least $.9 \times 2^{r(n)}$ or at most $.55 \times 2^{r(n)}$. Thus even a 10% approximate answer suffices to distinguish between the cases $x \in L$ and $x \notin L$. Thus approximating MAX PCP to within 10% suffices to decide membership for any language in NP. □

Theorem 5.3.1. *For every MAX SNP-hard problem Π, there exists a constant ϵ such that finding ϵ-approximate solutions to Π is NP-hard.*

Proof: The proof follows from the fact that there exists an approximation preserving reduction from MAX PCP to any MAX SNP-hard problem. In particular, given any MAX SNP-hard Π, there exists an ϵ such that an ϵ-approximate solution to Π would yield a 10% approximate solution to MAX PCP. Thus finding ϵ-approximate solutions to Π is NP-hard. □

5.4 Discussion

Addendum. The improvement in PCP technology has resulted in new hardness results for several optimization problems, as well as stronger hardness results for the ones discussed in this chapter. Most notable of these are the hardness of approximation results for Chromatic Number and Set Cover, due to Lund and Yannakakis [89], the hardness of approximating many maximal subgraph problems, also due to Lund and Yannakakis [90], the hardness of 4-coloring and 3-colorable graph, due to Khanna, Linial and Safra [82] and the hardness of the nearest vector and nearest lattice point problems, due to Arora, Babai, Stern and Sweedyk [7]. In almost all of these cases, hardness of approximation to within constant factors was known or could be inferred from the prior results [58, 8] – but the quality of the non-approximability results were far from the right answers. A more complete list of such hardness results may be found in the survey of Crescenzi and Kann [42]. These new results result in a much better understanding of the hardness results, at least qualitatively. For problems such as MAX 3 SAT, MAX 2 SAT, VERTEX COVER, MAXCUT, MAX CLIQUE, the qualitative level of approximability was resolved in [50, 6, 8]. However the actual constant upto which any of these problems can be approximated, or shown hard to approximate, is still a subject of active research. Starting with the work of Bellare, Goldwasser, Lund and Russell [28] and continuing through the works of Feige and Kilian [47], Bellare and Sudan [25] and Bellare, Goldreich and Sudan [27], the hardness results for all these problems have improved steadily and the best known hardness results at the time of this writeup are:

MAX 3 SAT 1/38-approximation is NP-hard [28, 47, 25, 27].

MAX 2 SAT 1/97-approximation is NP-hard. [27].

VERTEX COVER 1/26 approximation is NP-hard. [27].

MAXCUT 1/82 approximation is NP-hard. [27].

SET COVER $(1-\epsilon)\ln n$-approximation is hard, for every $\epsilon > 0$, unless NP $\subset$ RTIME($n^{\log\log n}$) [89, 28, 46].

MAX CLIQUE $n^{1/3-\epsilon}$ approximation is hard, for every $\epsilon > 0$, unless NP=RP [28, 47, 25, 27].

CHROMATIC NUMBER $n^{1/5-\epsilon}$ approximation is hard, for every $\epsilon > 0$, unless NP=RP [89, 82, 28, 47, 25, 57, 27].

6. Conclusions

We have proved that any NP language admits an efficient probabilistically checkable proof of membership. This proof need only be examined in a constant number of randomly chosen places by a polynomial time verifier. Currently, the transformation from the standard proof to the transparent proof requires a slightly super-quadratic blowup in size. Can this be substantially improved? We should point out that since mathematicians rarely write up proofs in enough detail to be machine checkable, our results should not be regarded as having practical consequences to mathematical proof checking. Nevertheless, it is possible that these techniques might be useful in ensuring software reliability – by making it possible to build redundancy into computations so that they can be efficiently checked. The blowup in the size of the above transformation is quite crucial for this application.

One interesting consequence of our results on testing and correcting of polynomials is that we obtain efficient randomized algorithms for error-detection and error-correction of some classical codes, like the Hadamard codes and the Reed-Solomon codes. The error-detection algorithm can very efficiently and by making very few probes into a received word, approximate its distance from a valid codeword. The error-correction algorithm can retrieve any bit of the nearest codeword to the received word by making very few probes into the received word. These error-detection and error-correction schemes have already been put to theoretical use in this work. It remains to be seen if they be put to practical use. The efficient randomized algorithms can be converted into fast deterministic parallel algorithms, and such error detecting and correcting schemes might be of some interest.

The surprising connection between efficient probabilistically checkable proofs and the hardness of approximating clique sizes in graphs, due to Feige, Goldwasser, Lovasz, Safra and Szegedy [50], is now much better understood. Here we showed that for every MAX SNP-hard problem, there exists a constant ϵ such that approximating the optimum value to within a relative error of ϵ is NP-hard. More recently, Lund and Yannakakis [89] have shown strong hardness results for approximating the chromatic number of graphs and approximating the minimum set cover size for a family of sets. In the light of all these developments, it appears that this connection between probabilistically checkable proofs and approximation problems is a fundamental one. Several questions still need to be resolved including the complexity of approximating the traveling salesman problem on the *plane*, approximating the longest path in a graph, finding the magnification of a graph, the length of shortest vector in a lattice etc. The last of these problems is particularly interesting since the approximate version of this problem is not very sensitive to the exact norm being used, and thus a hardness under any one norm would yield a hardness result under all norms. This is an example of one situation where the hardness of approximation might end up providing the first NP-completeness for even the exact problem, since the "shortest vector in a lattice" problem is not known to be NP-hard under arbitrary norms. A concrete example of such a result may be found in Lund and Yannakakis [89], where they show the hardness of approximating a certain problem thus providing the first proof showing hardness of exact computation for that problem.

The exact constant for the number of bits examined by probabilistic proof systems for NP is also very important, since this is directly related to the constants for which the different approximation problems become NP-hard. Of course, the number of bits examined can be traded off against the probability of discovering a fallacious proof. In fact, if one is willing to accept only a tiny probability of detecting a false proof, then for a certain proof system examining 3 bits of a proof is sufficient. Thus the precise version of the question would ask: How many bits of a proof need to be examined if the verifier is expected to reject false claims with probability half? At present, the best known bound on the number of bits probed in polynomial sized proofs seems to be less than a 100 bits Phillips and Safra [96]. Such a result would translate to showing that MAX 3 SAT cannot be approximated to within $\frac{1}{300}$. This, of course, is far from being tight (the best known upper bound for the constant is 1/8).

One way of interpreting the hardness result for approximating MAX 3 SAT is the following: we need to perturb the input in a large number of places to go from "yes" instances of a problem (one that is largely satisfiable) to "no" instances of a problem (one in which a significant fraction of the clauses are not satisfied under any assignment). Thus the hardness of this problem is not attributable to the sensitivity to the input. An extreme form of this interpretation leads us to consider the question of whether we can draw the input from the uniform distribution (or any polynomial time sampleable distribution) and still get problems that are very hard? Interestingly enough one of the key techniques relying on the power of polynomials – random self-reducibility – has already been used to show instances of average case hard problems for #P by Lipton [87]. It would be interesting to see if any of these techniques can be used to show average case hardness for problems in NP.

Addendum. The enormous amount of research that has one on in the areas of non-approximability and proof verification, has led to many of the questions raised above being already resolved. Polishchuk and Spielman [97] have constructed "nearly-linear" sized transparent proofs which are verifiable with constant probes. Arora, Babai, Stern and Sweedyk [7] have also brought much better understanding to the problem of approximating the shortest vector problem, though the complexity of the problem still remains open under most norms. The value of ϵ for which ϵ-approximating MAX 3 SAT (or MAX 2 SAT) is hard is getting closer to the value of ϵ for which ϵ-approximating MAX 3 SAT (MAX 2 SAT resp.) is achievable in polynomial time. This tightening of the thresholds is a result of both improved hardness results (see discussion at end of Chapter 5) as well as the discovery of new techniques for finding approximate solutions to optimization problems and in particular the use of semidefinite programming towards this end (see the work of Goemans and Williamson [64]). Currently these threshold are within a factor of 10 from each other and the gap may well narrow further in the near future.

Bibliography

1. M. Abadi, J. Feigenbaum, and J. Kilian. On hiding information from an oracle. *Journal of Computer and System Sciences*, 39:21–50, 1989.
2. L. Adleman, M. Huang, and K. Kompella. Efficient checkers for number-theoretic computations. *Information and Computation*, 121:1, 93–102, 1995.
3. S. Ar, R. Lipton, R. Rubinfeld, and M. Sudan. Reconstructing algebraic functions from mixed data. *Proceedings of the Thirty Third Annual Symposium on the Foundations of Computer Science*, IEEE, 1992.
4. S. Arora. *Probabilistic Checking of Proofs and Hardness of Approximation Problems*. PhD thesis, U.C. Berkeley, 1994.
5. S. Arora. Reductions, Codes, PCPs and Inapproximability. *Proceedings of the Thirty Sixth Annual Symposium on the Foundations of Computer Science*, IEEE, 1995.
6. S. Arora and S. Safra. Probabilistic checking of proofs: A new characterization of NP. *Proceedings of the Thirty Third Annual Symposium on the Foundations of Computer Science*, IEEE, 1992.
7. S. Arora, L. Babai, J. Stern and Z. Sweedyk. The hardness of approximating problems defined by linear constraints. *Proceedings of the Thirty Fourth Annual Symposium on the Foundations of Computer Science*, IEEE, 1993.
8. S. Arora, C. Lund, R. Motwani, M. Sudan, and M. Szegedy. Proof verification and the intractability of approximation problems. *Proceedings of the Thirty Third Annual Symposium on the Foundations of Computer Science*, IEEE, 1992.
9. S. Arora, R. Motwani, S. Safra, M. Sudan, and M. Szegedy. PCP and approximation problems. *Unpublished note*, 1992.
10. G. Ausiello, A. D'Atri, and M. Protasi. On the structure of combinatorial problems and structure preserving reductions. In *Proceedings of the 4th International Colloquium on Automata, Languages and Programming*, pages 45–57, 1977.
11. G. Ausiello, A. D'Atri, and M. Protasi. Structure preserving reductions among convex optimization problems. *Journal of Computer and Systems Sciences*, 21:136–153, 1980.
12. G. Ausiello, A. Marchetti-Spaccamela, and M. Protasi. Towards a unified approach for the classification of np-complete optimization problems. *Theoretical Computer Science*, 12:83–96, 1980.
13. L. Babai. Trading group theory for randomness. *Proceedings of the Seventeenth Annual Symposium on the Theory of Computing*, ACM, 1985.
14. L. Babai. Transparent (holographic) proofs. *Proceedings of the Tenth Annual Symposium on Theoretical Aspects of Computer Science*, Lecture Notes in Computer Science Vol. 665, Springer Verlag, 1993.

15. L. Babai. Transparent proofs and limits to approximation. *Proceedings of the First European Congress of Mathematics (1992)*, Vol. I, Birkhäuser Verlag, 1994, pp. 31–91.
16. L. Babai and L. Fortnow. Arithmetization: A new method in structural complexity theory. *Computational Complexity*, 1:41–66, 1991.
17. L. Babai and K. Friedl. On slightly superlinear transparent proofs. *Univ. Chicago Tech. Report*, CS-93-13, 1993.
18. L. Babai and S. Moran. Arthur-Merlin games: A randomized proof system and a hierarchy of complexity classes. *Journal of Computer and System Sciences*, 36:254–276, 1988.
19. L. Babai, L. Fortnow, L. Levin, and M. Szegedy. Checking computations in polylogarithmic time. *Proceedings of the Twenty Third Annual Symposium on the Theory of Computing*, ACM, 1991.
20. L. Babai, L. Fortnow, and C. Lund. Non-deterministic exponential time has two-prover interactive protocols. *Computational Complexity*, 1:3–40, 1991.
21. D. Beaver and J. Feigenbaum. Hiding instances in multioracle queries. *Proceedings of the Seventh Annual Symposium on Theoretical Aspects of Computer Science*, Lecture Notes in Computer Science Vol. 415, Springer Verlag, 1990.
22. D. Beaver, J. Feigenbaum, J. Kilian, and P. Rogaway. Security with low communication overhead. In *Proceedings of Crypto '90, Springer Verlag LNCS 537*, pages 62–76, 1991.
23. M. Bellare. Interactive proofs and approximation: reductions from two provers in one round. *Proceedings of the Second Israel Symposium on Theory and Computing Systems*, 1993.
24. M. Bellare and P. Rogaway. The complexity of approximating a nonlinear program. *Complexity of Numerical Optimization*, Ed. P.M. Pardalos, World Scientific (1993).
25. M. Bellare and M. Sudan. Improved non-approximability results. *Proceedings of the Twenty Sixth Annual Symposium on the Theory of Computing*, ACM, 1994.
26. M. Bellare, D. Coppersmith, J. Håstad, M. Kiwi and M. Sudan. Linearity testing in characteristic two. *Proceedings of the Thirty Sixth Annual Symposium on the Foundations of Computer Science*, IEEE, 1995.
27. M. Bellare, O. Goldreich and M. Sudan. Free bits, PCP and Inapproximability: Towards tight results. *Proceedings of the Thirty Sixth Annual Symposium on the Foundations of Computer Science*, IEEE, 1995. Full version avaliable from the Electronic Colloquium on Computational Complexity, http://www.eccc.uni-trier.de/eccc/.
28. M. Bellare, S. Goldwasser, C. Lund, and A. Russell. Efficient probabilistically checkable proofs. *Proceedings of the Twenty Fifth Annual Symposium on the Theory of Computing*, ACM, 1993. (See also Errata sheet in *Proceedings of the Twenty Sixth Annual Symposium on the Theory of Computing*, ACM, 1994).
29. M. Ben-Or, S. Goldwasser, J. Kilian, and A. Wigderson. Multi-prover interactive proofs: How to remove intractability assumptions. *Proceedings of the Twentieth Annual Symposium on the Theory of Computing*, ACM, 1988.
30. E. Berlekamp and L. Welch. Error correction of algebraic block codes. US Patent Number 4,633,470, 1986.
31. P. Berman and G. Schnitger. On the complexity of approximating the independent set problem. *Information and Computation*, 96:77–94, 1992.
32. M. Bern and P. Plassman. The steiner problem with edge lengths 1 and 2. *Information Processing Letters*, 32:171–176, 1989.

33. A. Blum, T. Jiang, M. Li, J. Tromp, and M. Yannakakis. Linear approximation of shortest superstrings. *Proceedings of the Twenty Third Annual Symposium on the Theory of Computing*, ACM, 1991.

34. M. Blum. Designing programs to check their work. Technical Report TR–88–009, International Computer Science Institute, 1988.

35. M. Blum and S. Kannan. Program correctness checking ... and the design of programs that check their work. *Proceedings of the Twenty First Annual Symposium on the Theory of Computing*, ACM, 1989.

36. M. Blum, M. Luby, and R. Rubinfeld. Self-testing/correcting with applications to numerical problems. *Journal of Computer and System Sciences*, 47:3, 1993.

37. J.Y. Cai, A. Condon, and R. Lipton. PSPACE is Provable by Two Provers in One Round. *Proceedings of the Sixth Annual Conference on Structure in Complexity Theory*, IEEE, 1991.

38. A. Cohen and A. Wigderson. Dispersers, deterministic amplification, and weak random sources. *Proceedings of the Thirtieth Annual Symposium on the Foundations of Computer Science*, IEEE, 1989.

39. A. Condon. The complexity of the max word problem, or the power of one-way interactive proof systems. *Proceedings of the Eighth Annual Symposium on Theoretical Aspects of Computer Science*, Lecture Notes in Computer Science Vol. 480, Springer Verlag, 1991.

40. A. Condon, J. Feigenbaum, C. Lund and P. Shor. Probabilistically Checkable Debate Systems and Approximation Algorithms for PSPACE-Hard Functions. *Proceedings of the Twenty Fifth Annual Symposium on the Theory of Computing*, ACM, 1993.

41. S.A. Cook. The complexity of theorem proving procedures. *Proceedings of the Third Annual Symposium on the Theory of Computing*, ACM, 1971.

42. P. Crescenzi and V. Kann. A compendium of NP optimization problems. Manuscript avalible from the authors, piluc@dsi.uniroma1.it or viggo@nada.kth.se.

43. E. Dalhaus, D.S. Johnson, C.H. Papadimitriou, P.D. Seymour, and M. Yannakakis. The complexity of multiway cuts. *SIAM Journal on Computing*, 23:4, pp. 864–894, 1994.

44. W.F. De la Vega and G.S. Lueker. Bin Packing can be solved within $1 + \epsilon$ in Linear Time. *Combinatorica*, vol. 1, pages 349–355, 1981.

45. R. Fagin. Generalized first-order spectra and polynomial-time recognizable sets. In R. M. Karp, editor, *Complexity of Computation, SIAM-AMS Proceedings, Vol. 7*, pages 43–73, 1974.

46. U. Feige. A threshold of $\ln n$ for approximating set cover. Manuscript, 1995.

47. U. Feige and J. Kilian. Two prover protocols – Low error at affordable rates. *Proceedings of the Twenty Sixth Annual Symposium on the Theory of Computing*, ACM, 1994.

48. U. Feige and L. Lovasz. Two-prover one-round proof systems: Their power and their problems. *Proceedings of the Twenty Fourth Annual Symposium on the Theory of Computing*, ACM, 1992.

49. U. Feige and C. Lund. On the hardness of computing the permanent of random matrices. *Proceedings of the Twenty Fourth Annual Symposium on the Theory of Computing*, ACM, 1992.

50. U. Feige, S. Goldwasser, L. Lovasz, S. Safra, and M. Szegedy. Approximating clique is almost NP-complete. *Proceedings of the Thirty Second Annual Symposium on the Foundations of Computer Science*, IEEE, 1991.

51. J. Feigenbaum. Locally random reductions in interactive complexity theory. In J. y. Cai, editor, *Complexity Theory, DIMACS Series on Discrete Mathematics and Theoretical Computer Science*, 1993.

52. J. Feigenbaum and L. Fortnow. On the random self-reducibility of complete sets. *Proceedings of the Sixth Annual Conference on Structure in Complexity Theory*, IEEE, 1991.
53. L. Fortnow, J. Rompel, and M. Sipser. On the power of multi-prover interactive protocols. *Proceedings of the Third Annual Conference on Structure in Complexity Theory*, IEEE, 1988.
54. K. Friedl, Zs. Hátsági and A. Shen. Low-degree testing. *Proceedings of the Fifth Symposium on Discrete Algorithms*, ACM, 1994.
55. K. Friedl and M. Sudan. Some improvements to low-degree tests. *Proceedings of the Third Israel Symposium on Theory and Computing Systems*, 1995.
56. R. Freivalds. Fast probabilistic algorithms. In *Lecture Notes in Computer Science*, pages 57–69. Springer-Verlag, 1979.
57. M. Furer. Improved hardness results for approximating the chromatic number. *Proceedings of the Thirty Sixth Annual Symposium on the Foundations of Computer Science*, IEEE, 1995.
58. M.R. Garey and D.S. Johnson. The complexity of near-optimal graph coloring. *Journal of the ACM*, 23:43–49, 1976.
59. M.R. Garey and D.S. Johnson. "strong" NP-completeness: Motivation, examples and implications. *Journal of the ACM*, 25:499–508, 1978.
60. M.R. Garey and D.S. Johnson. *Computers and Intractability: A Guide to the Theory of NP-Completeness*. W. H. Freeman, 1979.
61. M. Garey, D. Johnson and L. Stockmeyer. Some simplified NP-complete graph problems. *Theoretical Computer Science* 1, pp. 237–267, 1976.
62. P. Gemmell and M. Sudan. Highly resilient correctors for polynomials. *Information Processing Letters* 43 (1992), 169–174.
63. P. Gemmell, R. Lipton, R. Rubinfeld, M. Sudan, and A. Wigderson. Self-testing/correcting for polynomials and for approximate functions. *Proceedings of the Twenty Third Annual Symposium on the Theory of Computing*, ACM, 1991.
64. M. Goemans and D. Williamson. .878 approximation algorithms for Max-CUT and Max-2SAT. *Proceedings of the Twenty Sixth Annual Symposium on the Theory of Computing*, ACM, 1994.
65. O. Goldreich. A taxonomy of proof systems. In SIGACT News complexity theory column 4, *SIGACT News* Vol. 25, No. 1, 1994.
66. O. Goldreich. Probabilistic proof systems. *Proceedings of the International Congress of Mathematicians*, 1994.
67. O. Goldreich and L. Levin. A hard-core predicate for any one-way function. *Proceedings of the Twenty First Annual Symposium on the Theory of Computing*, ACM, 1989.
68. O. Goldreich, S. Micali, and A. Wigderson. Proofs that yield nothing but their validity and a methodology of cryptographic protocol design. *Proceedings of the Twenty Seventh Annual Symposium on the Foundations of Computer Science*, IEEE, 1986.
69. O. Goldreich, R. Rubinfeld and M. Sudan. Learning polynomials with queries: The highly noisy case. *Proceedings of the Thirty Sixth Annual Symposium on the Foundations of Computer Science*, IEEE, 1995.
70. S. Goldwasser, S. Micali, and C. Rackoff. The knowledge complexity of interactive proof systems. *SIAM Journal on Computing*, 18:186–208, 1989.
71. D.S. Hochbaum and D.B. Shmoys. A best possible heuristic for the k-center problem. *Mathematics of Operations Research*, 10(2):180–184, May 1985.
72. D.S. Hochbaum and D.B. Shmoys. A unified approach to approximation algorithms for bottleneck problems. *Journal of the ACM*, 33(3):533–550, July 1986.

73. D.S. Hochbaum and D.B. Shmoys. Using dual approximation algorithms for scheduling problems: Theoretical and practical results. *Journal of the ACM*, 34(1):144–162, January 1987.
74. R. Impagliazzo and D. Zuckerman. How to recycle random bits. *Proceedings of the Thirtieth Annual Symposium on the Foundations of Computer Science*, IEEE, 1989.
75. D. Johnson. Approximation algorithms for combinatorial problems. *Journal of Computer and Systems Sciences*, 9 (1974), 256–278.
76. D. Johnson. The NP-completeness column: an ongoing guide. *Journal of Algorithms*, 13, 1992.
77. V. Kann. Maximum bounded 3-dimensional matching is MAX SNP-complete. *Information Processing Letters*, 37:27–35, 1991.
78. S. Kannan. *Program Result Checking with Applications*. PhD thesis, University of California, Berkeley, 1990.
79. D. Karger, R. Motwani, and G.D.S. Ramkumar. On approximating the longest path in a graph. *Proceedings Workshop on Algorithms and Data Structures*, Lecture Notes in Computer Science (Springer-Verlag), vol. 709, pp. 421–430, 1993.
80. R.M. Karp. Reducibility among combinatorial problems. In R.E. Miller and J.W. Thatcher, editors, *Complexity of Computer Computations*, pages 85–103. Plenum Press, 1972.
81. N. Karmakar and R.M. Karp. An Efficient Approximation Scheme For The One-Dimensional Bin Packing Problem. *Proceedings of the Twenty Third Annual Symposium on the Foundations of Computer Science*, IEEE, 1982.
82. S. Khanna, N. Linial and S. Safra. On the hardness of approximating the chromatic number. *Proceedings of the Second Israel Symposium on Theory and Computing Systems*, 1993.
83. S. Khanna, R. Motwani, M. Sudan and U. Vazirani. On syntactic versus computational views of approximability. *Proceedings of the Thirty Fifth Annual Symposium on the Foundations of Computer Science*, IEEE, 1994.
84. P. Kolaitis and M. Vardi. The decision problem for the probabilities of higher-order properties. *Proceedings of the Nineteenth Annual Symposium on the Theory of Computing*, ACM, 1987.
85. D. Lapidot and A. Shamir. Fully parallelized multi prover protocols for NEX-PTIME. *Proceedings of the Thirty Second Annual Symposium on the Foundations of Computer Science*, IEEE, 1991.
86. L. Levin. Universal'nyĭe perebornyĭe zadachi (Universal search problems, in Russian). *Problemy Peredachi Informatsii*, 9:265–266, 1973. A corrected English translation appears in an appendix to Trakhtenbrot [109].
87. R. Lipton. New directions in testing. In J. Feigenbaum and M. Merritt, editors, *Distributed Computing and Cryptography, DIMACS Series in Discrete Math and Theoretical Computer Science, American Mathematical Society*, 2:191–202, 1991.
88. M. Luby. A simple parallel algorithm for the maximal independent set problem. *SIAM Journal of Computing*, 15(4):1036–1053, 1986.
89. C. Lund and M. Yannakakis. On the hardness of approximating minimization problems. Preprint, June 1992.
90. C. Lund and M. Yannakakis. The approximation of maximum subgraph problems. *Proceedings of ICALP 93*, Lecture Notes in Computer Science Vol. 700, Springer Verlag, 1993.
91. C. Lund, L. Fortnow, H. Karloff, and N. Nisan. Algebraic methods for interactive proof systems. *Proceedings of the Thirty First Annual Symposium on the Foundations of Computer Science*, IEEE, 1990.

92. R. Motwani. Lecture Notes on Approximation Algorithms. Technical Report, Dept. of Computer Science, Stanford University (1992).

93. C. Papadimitriou and M. Yannakakis. Optimization, approximation and complexity classes. *Journal of Computer and System Sciences*, 43:425–440, 1991.

94. C. Papadimitriou and M. Yannakakis. The traveling salesman problem with distances one and two. Journal of *Mathematics of Operations Research*, 1992.

95. A. Paz and S. Moran. Non-deterministic polynomial optimization problems and their approximation. *Theoretical Computer Science*, 15:251–277, 1981.

96. S. Phillips and S. Safra. PCP and tighter bounds for approximating MAXSNP. *Manuscript*, Stanford University, 1992.

97. A. Polishchuk and D. Spielman. Nearly Linear Sized Holographic Proofs. *Proceedings of the Twenty Sixth Annual Symposium on the Theory of Computing*, ACM, 1994.

98. R. Raz. A parallel repetition theorem. *Proceedings of the Twenty Seventh Annual Symposium on the Theory of Computing*, ACM, 1995.

99. I.S. Reed and G. Solomon. Polynomial codes over certain finite fields. *Journal of the Society of Industrial and Applied Mathematics*, 8:300–304, June 1960.

100. R. Rubinfeld. *A Mathematical Theory of Self-Checking, Self-Testing and Self-Correcting Programs*. PhD thesis, University of California at Berkeley, 1990.

101. R. Rubinfeld and M. Sudan. Testing polynomial functions efficiently and over rational domains. *Proceedings of the Third Symposium on Discrete Algorithms*, ACM, 1994.

102. R. Rubinfeld and M. Sudan. Robust characterizations of polynomials and their applications to program testing. To appear in *SIAM Journal on Computing*. Technical Report RC 19156, IBM Research Division, T. J. Watson Research Center, P.O. Box 218, Yorktown Heights, NY 10598, September 1993.

103. S. Sahni and T. Gonzalez. P-complete approximation problems. *Journal of the ACM*, 23:555–565, 1976.

104. J. Schwartz. Fast probabilistic algorithms for verification of polynomial identities. *Journal of the ACM*, 27:701–717, 1980.

105. A. Shamir. IP = PSPACE. In *Proceedings of the 31st Annual IEEE Symposium on Foundations of Computer Science*, pages 11–15, 1990.

106. A. Shen. Personal Communication, May 1991.

107. D. Shmoys. Computing near-optimal solutions to combinatorial optimization problems. In *W. Cook, L. Lovasz, and P.D. Seymour, Editors*, DIMACS volume on Combinatorial Optimization, 1995.

108. G. Tardos. Multi-prover encoding schemes and three prover proof systems. *Proceedings of the Ninth Annual Conference on Structure in Complexity Theory*, IEEE, 1994.

109. B.A. Trakhtenbrot. A survey of Russian approaches to *perebor* (brute-force search) algorithms. *Annals of the History of Computing*, 6:384–400, 1984.

110. van der Waerden. *Algebra*, volume 1. Frederick Ungar Publishing Co., Inc., 1970.

111. M. Yannakakis. On the approximation of maximum satisfiability. *Proceedings of the Third Symposium on Discrete Algorithms*, ACM, 1994.

112. D. Zuckerman. NP-Complete Problems have a version that is hard to Approximate. *Proceedings of the Eighth Annual Conference on Structure in Complexity Theory*, IEEE, 1993.

A. The Berlekamp Welch decoder

The essence of the Berlekamp Welch technique lies in their ability to use rational functions to describe a sequence of points most of which lie on a univariate polynomial. We first give some basic facts about rational functions. In this section, we can allow these functions to be functions over any field F.

A.1 Preliminaries: rational functions

Definition A.1.1 (rational function). *A function $r : F \to F$ is a rational function if it can be expressed as $r(x) = \frac{f(x)}{g(x)}$ for polynomials f and g. The* **degree** *of $r(x)$ is given by the ordered pair (d_1, d_2), where d_1 is the degree of $f(x)$ and d_2 is the degree of $g(x)$.*

Definition A.1.2. *A rational function $r(x)$ describes a sequence $\{(x_i, y_i)| i = 1 \text{ to } n\}$ if for all i, $y_i = r(x_i)$ or $r(x_i)$ evaluates to $0/0$.*

Fact A.1.1 (uniqueness). If rational functions $\frac{f_1(x)}{g_1(x)}$ and $\frac{f_2(x)}{g_2(x)}$ both of degree (d_1, d_2) where $d_1 + d_2 < n$ describe a given sequence $\{(x_i, y_i)|i = 1 \text{ to } n\}$, then $\frac{f_1}{g_1} \equiv \frac{f_2}{g_2}$.

Proof [Sketch]: Follows from the fact that the polynomials $f_1 * g_2$ and $f_2 * g_1$ agree at each of the points x_i. But these are both polynomials of degree $d_1 + d_2$ and if they agree at $n > d_1 + d_2$ points, then they must be identical. Hence $\frac{f_1}{g_1} \equiv \frac{f_2}{g_2}$. $\qquad\square$

Fact A.1.2 (interpolation). Given a sequence $\{(x_i, y_i)|i = 1 \text{ to } n\}$, a rational function of degree (d_1, d_2) that describes the given sequence can be found in time polynomial in n, provided one exists.

Proof: Observe that if we let the coefficients of f and g be unknowns, then the constraints $f(x_i) = y_i * g(x_i)$ become linear constraints in the unknowns. Thus if the linear system so obtained has a solution, then it can be found by matrix inversion. $\qquad\square$

A.2 The decoder

Recall that the task we wish to solve is the following:
Given: *n points* $\{(x_i, y_i) | i = 1 \ to \ n\}$.
Output: *A polynomial p of degree at most d such that for all but k values of i, $y_i = p(x_i)$ (where $2k + d < n$).*

Claim. There exists a rational function r of degree $(k + d, k)$ which describes the given sequence.

Proof: Consider a polynomial W which evaluates to zero at x_i if $y_i \neq p(x_i)$. It is clear that such a polynomial with degree is at most k exists. Now consider the rational function $\frac{p \cdot W}{W}$. This describes all the input points.   □
  We are now in a position to prove Lemma 2.3.1:
Lemma 2.3.1 Given n points $(x_i, y_i) \in F^2$, there exists an algorithm which finds a degree d polynomial g such that $g(x_i) = y_i$ for all but k values of i, where $2k + d < n$, if such a g exists. The running time of the algorithm is polynomial in d and n.
Proof: Claim A.2 tells us that there exists a rational function of the form $\frac{g \cdot W}{W}$ which describes the given points. A rational function which describes the given points can be found by interpolation and the rational functions are unique except for multiplication by common factors, and thus are of the form $\frac{g \cdot W'}{W'}$. Thus the quotient of the so obtained rational function gives us g.   □

B. Composing proof systems

Here we prove Lemma 4.3.1. Recall the definition of rPCP. We wish to prove:

Lemma 4.3.1 If NP $\subset$ rPCP$(r_1(n), q_1(n))$ and NP $\subset$ rPCP$(r_2(n), q_2(n))$, then NP $\subset$ rPCP$(r_1(n) + r_2(q_1(n)), q_2(q_1(n)))$.

Proof: The proof follows in a straightforward manner based on the discussion of Section 4.3. The proof is somewhat long since we have to ensure that the composed system satisfies all the properties required of an rPCP proof system.

Input Encoding. Let E_1 be the coding scheme for the rPCP$(r_1(n), q_1(n))$ proof system and let E_2 be the coding scheme for the rPCP$(r_2(n), q_2(n))$ proof system. Then the coding E for the composed proof system is obtained by first encoding x according to E_1 and thus obtaining a $s_1(n) \times q_1(n)$ table T_{1x} and then encoding each $q_1(n)$-bit entry of T_{1x} as a $s_2(q_1(n)) \times q_2(q_1(n))$ entry. The final table for x thus contains $s_1(n) \times s_2(q_1(n))$ entries each of which is $q_2(q_1(n))$ bits long.

The inverse mapping E^{-1} is obtained by viewing a table $T : [s_1(n)] \times [s_2(q_1(n))] \to \{0, 1\}^{q_2(q_1(n))}$ as $s_1(n)$ tables of size $s_2(q_1(n)) \times q_2(q_1(n))$ and inverting each of these tables according to E_2^{-1}. This gives $s_1(n)$ entries of size $q_1(n)$ which can be viewed as a $s_1(n) \times q_1(n)$ table which when inverted according to E_1^{-1} gives an n-bit string. This n bit string is defined to the inverse according to E^{-1} of the table T.

Proof Tables. Let π_{1x} be the proof table for $x \in L$ according to the rPCP$(r_1(n), q_1(n))$ proof system. Then the first portion of the proof table π for the composed system consists of $s_1(n)$ tables of size $s_2(q_1(n)) \times q_2(q_1(n))$. These are the tables obtained by encoding each entry of π_{1x} according to E_2.

The second portion of π consists of one table for each possible random string $r \in \{0, 1\}^{r_1(n)}$. Let $y_1, \cdots, y_{c'}$ be the contents of the c' locations of the tables $E_1(x_1), \ldots, E_1(x_c)$ and π_1 that are read by the tester T_1 on the choice of r as a random string. Further, let C_r be the circuit of size poly$(q_1(n))$ which decides whether to accept $y_1, \cdots, y_{c'}$ or not. Then the table π contains a table π_{2r} of size $s_2(q_1(n)) \times q_2(q_1(n))$ which proves that $(C_r, y_1, \ldots, y_{c'})$ represents a circuit with an accepting input assignment in $y_1 \cdots y_{c'}$. (Notice that this is a polytime computable predicate and hence in NP.)

Tester. The tester T for the composed proof system acts as follows. It first picks a random string $r \in \{0,1\}^{r_1(n)}$ and tries to simulate the action of the tester T_1. This would involve reading $y_1, \ldots, y_{c'}$ and then verifying that a circuit C_r will accept $(y_1, \ldots, y_{c'})$. The tester will thus encode the representation of the circuit C_r using E_2 and then use the second proof system and the proof π_{2r} to verify that $E_2(C_r), E_2(y_1), \ldots, E_2(y_{c'})$ represents a circuit with an accepting input assignment. Notice further that the computation of the tester T can be expressed as a circuit of size $\text{poly}(q_2(q_1(n)^{O(1)}))$ (which is the same circuit which describes the computation of tester T_2 on input $E_2(C_r), E_2(y_1), \ldots, E_2(y_{c'})$ and proof π_{2r}).

Correctness. It is clear that if $x_1 \cdot x_2 \cdots x_c \in L$ then there exists a proof such that the tester T always outputs PASS. For the other direction, consider tables $\tau_1, \ldots, \tau_c$ such that $E^{-1}(\tau_1) \cdots E^{-1}(\tau_c) \notin L$. Let π be a proof which tries to prove that $E^{-1}(\tau_1) \cdots E^{-1}(\tau_c) \in L$. We will show that the tester T will output FAIL on this proof with probability at least $1/4$.

Let $\tau_1^{(1)}, \ldots, \tau_c^{(1)}$ be the $s_1(n) \times q_1(n)$ tables obtained by interpreting the tables $\tau_1, \ldots, \tau_c$ as $s_1(n)$ tables of size $s_2(q_1(n)) \times q_2(q_1(n))$ and decoding each such table according to E_2^{-1}. Similarly let π_1 be the $s_1(n) \times q_1(n)$ table obtained by decoding the tables of the first half of π according to Then $E_1^{-1}(\tau_1^{(1)}) \cdot E_1^{-1}(\tau_1^{(c)}) \notin L$. Therefore we find that for half the choices of $r \in \{0,1\}^{r_1(n)}$, the tester $T_{1,r}(E_1^{-1}(\tau_1^{(1)}) \cdot E_1^{-1}(\tau_1^{(c)}), \pi_1)$ will output FAIL. Now for all such choices of r, Let the contents of the entries of $\tau_1, \ldots, \tau_c$ and π_1 as read by the tester T_1 be $y_1, \ldots, y_{c'}$. Then, $y_1, \ldots, y_{c'}$ are the inverse encodings of some tables $\tau_1^{(2)}, \ldots, tau_{c'}^{(2)}$ according to E_2^{-1}. $T_{1,r}(y_1, \ldots, y_{c'})$ outputs FAIL, By the property of the second proof system, we have that for at least half the choices of $r^{(2)} \in \{0,1\}^{r_2(q_1(n))}$, the tester $T_{2,r^{(2)}}(\tau_1^{(2)}, \ldots, tau_{c'}^{(2)}, \pi_2$ will output FAIL for any proof π_2. Thus with probability $1/4$ the tester T outputs fail on any proof π.

By running the same tester thrice on this proof system, the probability of outputting fail can be increased to $37/64$ which is greater than a half.

$\square$

C. A characterization of NP via polynomial sequences

We first show how to construct a sequence of polynomials which verifies that a 3-CNF formula is satisfiable.

Lemma 4.2.1 ([20, 19]) For a 3-CNF formula ϕ there exists a polynomial sequence of length and width at most $\log n$, with polynomials of degree at most $\log^2 n$ in $\Theta(\frac{\log n}{\log\log n})$ variables, such that sequence can be terminated in the trivial polynomial if and only if ϕ is satisfiable.

Proof: The proof places the formula ϕ and the assignment A in a suitable encoding and then verifies the consistency of A for ϕ.

To encode n-bits $a_1, \ldots, a_n$ which represent an assignment to n variables, we will use the (m, h, F)-polynomial extension encoding, where $h^m = n$. Recall that this means that we pick a set $H \subset F$ (with $|H| = h$) and let A be the function from H^m to $\{0, 1\}$ which is specified on n places by $a_1, \ldots, a_n$. The encoding is the extension of A to a low-degree polynomial $\tilde{A}$ (of degree at most mh).

The encoding of the 3-CNF formula is obtained as follows: We view the formula as a function f from $H^{3m} \times \{0, 1\}^3$ to $\{0, 1\}$, where $f(z_1, z_2, z_3, b_1, b_2, b_3) = 1$ if the formula has a clause which contains the variable z_1, z_2, z_3 with negated variables indicated by the bit b_i being 1. The encoding of this formula will be a function $\tilde{f} : F^{3m+3} \to F$.

Consider the function $\mathbf{sat?} : F^{3m+3} \to F$ defined as follows:

$$\mathbf{sat?}(z_1, z_2, z_3, b_1, b_2, b_3) = \prod_{i=1}^{3} (\tilde{A}(z_i) * b_i + (1 - \tilde{A}(z_i)) * (1 - b_i))$$

$\mathbf{sat?}$ is a polynomial of degree $O(mh)$ and evaluates to zero in the domain $H^{3m} \times \{0, 1\}^3$ if the clause given by $(z_1, z_2, z_3, b_1, b_2, b_3)$ is satisfied by the assignment.

Now consider the polynomial

$$g^{(0)}(z_1, z_2, z_3, b_1, b_2, b_3) = f(z_1, z_2, z_3, b_1, b_2, b_3) * \mathbf{sat?}(z_1, z_2, z_3, b_1, b_2, b_3)$$

Our task is to ensure that $g^{(0)}$ is zero on the domain $H^{3m} \times \{0, 1\}^3$.

Let $m' = 3m + 3$. $g^{(0)}$ is a function on m' variables, say $z_1, \ldots, z_{m'}$, and we wish to ensure that g restricted to $H^{m'}$ is zero. We do so by constructing a sequence of polynomials such that the final polynomial will not be zero if the initial one is not zero on this subdomain. *The intuition behind the rest of this* *proof is as follows. Effectively we want to simulate the effect of an OR gate which has fanin $H^{m'}$, by a polynomial sequence with small width and length. So first we implement the OR gate with fanin $H^{m'}$ by a circuit of depth m' and in which each gate has fanin H (in the obvious way). Then we show how to implement each OR gate in this circuit – approximately – by a low degree polynomial. This leads to a polynomial sequence whose width is the fanin of the circuit and whose length is the depth of the circuit.*

The polynomial $g^{(i)}$ is a function of the variables $z_{i+1}, \ldots, z_{m'}$ and $w_1, \ldots, w_i$ and is defined as follows:

$$g^{(i)}(z_{i+1}, ..., z_{m'}; w_1, ..., w_i) = \sum_{y \in H} g^{(i-1)}(y, z_{i+1}, \ldots, z_{m'}; w_1, \ldots, w_{i-1}) * w_i^{\zeta(y)}$$

(where ζ is some function that maps the elements in H to distinct integers in the range $0, \ldots, h - 1$).

It is clear that the $g^{(i)}$'s are polynomials of degree $O(h)$ in each variable. To complete the argument we wish to show that $g^{(m')}$ identically becomes zero if and only if $g^{(0)}$ is zero on the domain $H^{m'}$.

Let $r_1, \ldots, r_{m'}$ be randomly chosen from F. We will show that $g^{(i)}(z_{i+1}, \ldots, z_{m'}; r_1, \ldots, r_i)$ is non-zero with high probability, if $g^{(i-1)}(y, z_{i+1}, \ldots, z_{m'}; r_1, \ldots, r_{i-1})$ is non-zero for any $y \in H$. Let $c_y = g^{(i-1)}(y, z_{i+1}, \ldots, z_{m'}; r_1, \ldots, r_{i-1})$. Then $g^{(i)}(z_{i+1}, \ldots, z_{m'}; r_1, \ldots, w_i) = \sum_{y \in H} c_y * w_i^{\zeta(y)}$, is a univariate polynomial of degree h in w_i which is not identically zero. Thus for a random choice of $w_i = r_i$ the summation is non-zero with probability $1 - h/|F|$. Conversely, also observe that if the c_y's are all zero then the weighted sum is zero. This gives us the "approximate" OR gate we were looking for. By a simple argument, it can now be shown that with probability $1 - m'h/|F|$, $g^{(m')}(r_1, \ldots, r_{m'})$ is non-zero, implying that $g^{(m')}$ is not identically zero.

Thus we have shown how to build a construction sequence A,sat?, $g^{(0)}, \ldots, g^{(m')}$ with the property that $\exists A$ such that $g^{(m')} \equiv 0$ if and only if f represents a satisfiable 3-CNF formula.

Observe further that the length, width and degree of the sequence is as promised.   □

Now we are in a position to show a stronger version of this lemma: i.e., where the inputs are themselves part of the sequence. We will let the inputs $x_1, \ldots, x_c$ be encoded by the (m, h, F)-polynomial extension code $E_{m,h,F}$. The decoder $E_{m,h,F}^{-1}$ finds the closest polynomial to a function $f : F^m \to F$ and uses its values on the space H^m as the message. (Observe that the message so obtained is a message from $|F|^n$ and not necessarily $\{0, 1\}^n$. Since L will only include strings from $\{0, 1\}^n$, we will have to exclude all the remaining strings explicitly.)

Theorem 4.4.1 Given a constant $c > 0$ and language $L \in$ NP and a parameter n, a $(\frac{\log n}{\log \log n}, \log n)$-construction rule for degree $\log^2 n$ polynomials $g^{(0)}, \ldots, g^{(l)}$ where $g^{(i)} : F^m \to F$, can be constructed in polynomial time, with the property that $\exists g^{(0)}$ s.t. $g^{(l)} \equiv 0 \Leftrightarrow x_1 \cdot x_2 \cdots x_c \in L$, where

$x_i \in F^n$ is the decoding according to $E^{-1}_{m,h,F}$ of the polynomial $g^{(i)}$. Moreover $|F| = O(\text{polylog } n)$ and $m = \Theta(\frac{\log n}{\log\log n})$. [1]

Proof: The proof follows from the following basic ideas:

- (**Cook's Theorem**) Given c, n and L we can construct a 3-CNF formula ϕ_n on $n' = cn + n^{O(1)}$ such that $\exists y$ such that $\phi_n(x_1, x_2, \ldots, x_c, y)$ is true if and only if $x_1 \cdots x_c \in L$.
- (**Assignment**) Let $g^{(0)}$ be the $E_{m,h,F}$ encoding of the assignment to y. Let $g^{(i)}$ be the encodings of x_i. Construct $g^{(c+1)} : F^{m+1} \to F$ to be $g^{(c+1)}(i, z) = g^{(i)}(z)$. We will name this the assignment function A.

Now we are in a position to apply the lemma described above and construct a sequence of polynomials $g^{(c+2)}, \ldots, g^{(l)}$ such that the final polynomial is zero if and only if the assignment given by A satisfies ϕ_n.

□

---

[1] Some attributional remarks: Most of the technical work required for this theorem was done by Babai, Fortnow and Lund [20], but their characterization refers specifically only to NEXPTIME languages. The works of Babai, Fortnow, Levin and Szegedy [19] and Feige, Goldwasser, Lovasz, Safra and Szegedy [50] scale down the [20] proof to NP languages. In order to scale down this result Babai, Fortnow, Levin and Szegedy [19] needed more compressed representations of the assignment function than found in the encoding used by Babai, Fortnow and Lund [20]. They were the first to use the particular size of H as used in the proof here. The observation that one could work with a constant number of encoded inputs is made by Arora and Safra [6].

Index

Lecture Notes in Computer Science

For information about Vols. 1–954

please contact your bookseller or Springer-Verlag